PLUS: Official SQA Specimen Paper & 2015 Past Paper With Answers

Higher for CfE
Chemistry

2014 Specimen Question Paper, Model Papers & 2015 Exam

Hodder Gibson Study Skills Advice – General	– page 3
Hodder Gibson Study Skills Advice – Higher for CfE Chemistry	– page 5
2014 SPECIMEN QUESTION PAPER	– page 9
MODEL PAPER 1	– page 55
MODEL PAPER 2	– page 93
MODEL PAPER 3	– page 135
2015 EXAM	– page 175
ANSWER SECTION	– page 225

This book contains the official 2014 SQA Specimen Question Paper and 2015 Exam for Higher for CfE Chemistry, with associated SQA approved answers modified from the official marking instructions that accompany the paper.

In addition the book contains model papers, together with answers, plus study skills advice. These papers, some of which may include a limited number of previously published SQA questions, have been specially commissioned by Hodder Gibson, and have been written by experienced senior teachers and examiners in line with the new Higher for CfE syllabus and assessment outlines, Spring 2014. This is not SQA material but has been devised to provide further practice for Higher for CfE examinations in 2015 and beyond.

Hodder Gibson is grateful to the copyright holders, as credited on the final page of the Answer Section, for permission to use their material. Every effort has been made to trace the copyright holders and to obtain their permission for the use of copyright material. Hodder Gibson will be happy to receive information allowing us to rectify any error or omission in future editions.

Hachette UK's policy is to use papers that are natural, renewable and recyclable products and made from wood grown in sustainable forests. The logging and manufacturing processes are expected to conform to the environmental regulations of the country of origin.

Orders: please contact Bookpoint Ltd, 130 Park Drive, Milton Park, Abingdon, Oxon OX14 4SE. Telephone: (44) 01235 827720. Fax: (44) 01235 400454. Lines are open 9.00–5.00, Monday to Saturday, with a 24-hour message answering service. Visit our website at www.hoddereducation.co.uk. Hodder Gibson can be contacted direct on: Tel: 0141 848 1609; Fax: 0141 889 6315; email: hoddergibson@hodder.co.uk

This collection first published in 2015 by
Hodder Gibson, an imprint of Hodder Education,
An Hachette UK Company
2a Christie Street
Paisley PA1 1NB

CfE Higher Specimen Question Paper and Answers and CfE Higher Exam Paper 2015 and Answers © Scottish Qualifications Authority. Model Question Papers, Answers and Study Skills section © Hodder Gibson. All rights reserved. Apart from any use permitted under UK copyright law, no part of this publication may be reproduced or transmitted in any form or by any means, electronic or mechanical, including photocopying and recording, or held within any information storage and retrieval system, without permission in writing from the publisher or under licence from the Copyright Licensing Agency Limited. Further details of such licences (for reprographic reproduction) may be obtained from the Copyright Licensing Agency Limited, Saffron House, 6–10 Kirby Street, London EC1N 8TS.

Typeset by Aptara, Inc.

Printed in the UK

A catalogue record for this title is available from the British Library

ISBN: 978-1-4718-6071-3

3 2 1

2016 2015

Introduction

Study Skills – what you need to know to pass exams!

Pause for thought

Many students might skip quickly through a page like this. After all, we all know how to revise. Do you really though?

Think about this:

"IF YOU ALWAYS DO WHAT YOU ALWAYS DO, YOU WILL ALWAYS GET WHAT YOU HAVE ALWAYS GOT."

Do you like the grades you get? Do you want to do better? If you get full marks in your assessment, then that's great! Change nothing! This section is just to help you get that little bit better than you already are.

There are two main parts to the advice on offer here. The first part highlights fairly obvious things but which are also very important. The second part makes suggestions about revision that you might not have thought about but which WILL help you.

Part 1

DOH! It's so obvious but …

Start revising in good time

Don't leave it until the last minute – this will make you panic.

Make a revision timetable that sets out work time AND play time.

Sleep and eat!

Obvious really, and very helpful. Avoid arguments or stressful things too – even games that wind you up. You need to be fit, awake and focused!

Know your place!

Make sure you know exactly **WHEN and WHERE** your exams are.

Know your enemy!

Make sure you know what to expect in the exam.

How is the paper structured?

How much time is there for each question?

What types of question are involved?

Which topics seem to come up time and time again?

Which topics are your strongest and which are your weakest?

Are all topics compulsory or are there choices?

Learn by DOING!

There is no substitute for past papers and practice papers – they are simply essential! Tackling this collection of papers and answers is exactly the right thing to be doing as your exams approach.

Part 2

People learn in different ways. Some like low light, some bright. Some like early morning, some like evening / night. Some prefer warm, some prefer cold. But everyone uses their BRAIN and the brain works when it is active. Passive learning – sitting gazing at notes – is the most INEFFICIENT way to learn anything. Below you will find tips and ideas for making your revision more effective and maybe even more enjoyable. What follows gets your brain active, and active learning works!

Activity 1 – Stop and review

Step 1

When you have done no more than 5 minutes of revision reading STOP!

Step 2

Write a heading in your own words which sums up the topic you have been revising.

Step 3

Write a summary of what you have revised in no more than two sentences. Don't fool yourself by saying, "I know it, but I cannot put it into words". That just means you don't know it well enough. If you cannot write your summary, revise that section again, knowing that you must write a summary at the end of it. Many of you will have notebooks full of blue/black ink writing. Many of the pages will not be especially attractive or memorable so try to liven them up a bit with colour as you are reviewing and rewriting. **This is a great memory aid, and memory is the most important thing.**

Activity 2 – Use technology!

Why should everything be written down? Have you thought about "mental" maps, diagrams, cartoons and colour to help you learn? And rather than write down notes, why not record your revision material?

What about having a text message revision session with friends? Keep in touch with them to find out how and what they are revising and share ideas and questions.

Why not make a video diary where you tell the camera what you are doing, what you think you have learned and what you still have to do? No one has to see or hear it, but the process of having to organise your thoughts in a formal way to explain something is a very important learning practice.

Be sure to make use of electronic files. You could begin to summarise your class notes. Your typing might be slow, but it will get faster and the typed notes will be easier to read than the scribbles in your class notes. Try to add different fonts and colours to make your work stand out. You can easily Google relevant pictures, cartoons and diagrams which you can copy and paste to make your work more attractive and **MEMORABLE**.

Activity 3 – This is it. Do this and you will know lots!

Step 1

In this task you must be very honest with yourself! Find the SQA syllabus for your subject (www.sqa.org.uk). Look at how it is broken down into main topics called MANDATORY knowledge. That means stuff you MUST know.

Step 2

BEFORE you do ANY revision on this topic, write a list of everything that you already know about the subject. It might be quite a long list but you only need to write it once. It shows you all the information that is already in your long-term memory so you know what parts you do not need to revise!

Step 3

Pick a chapter or section from your book or revision notes. Choose a fairly large section or a whole chapter to get the most out of this activity.

With a buddy, use Skype, Facetime, Twitter or any other communication you have, to play the game "If this is the answer, what is the question?". For example, if you are revising Geography and the answer you provide is "meander", your buddy would have to make up a question like "What is the word that describes a feature of a river where it flows slowly and bends often from side to side?".

Make up 10 "answers" based on the content of the chapter or section you are using. Give this to your buddy to solve while you solve theirs.

Step 4

Construct a wordsearch of at least 10 × 10 squares. You can make it as big as you like but keep it realistic. Work together with a group of friends. Many apps allow you to make wordsearch puzzles online. The words and phrases can go in any direction and phrases can be split. Your puzzle must only contain facts linked to the topic you are revising. Your task is to find 10 bits of information to hide in your puzzle, but you must not repeat information that you used in Step 3. DO NOT show where the words are. Fill up empty squares with random letters. Remember to keep a note of where your answers are hidden but do not show your friends. When you have a complete puzzle, exchange it with a friend to solve each other's puzzle.

Step 5

Now make up 10 questions (not "answers" this time) based on the same chapter used in the previous two tasks. Again, you must find NEW information that you have not yet used. Now it's getting hard to find that new information! Again, give your questions to a friend to answer.

Step 6

As you have been doing the puzzles, your brain has been actively searching for new information. Now write a NEW LIST that contains only the new information you have discovered when doing the puzzles. Your new list is the one to look at repeatedly for short bursts over the next few days. Try to remember more and more of it without looking at it. After a few days, you should be able to add words from your second list to your first list as you increase the information in your long-term memory.

FINALLY! Be inspired...

Make a list of different revision ideas and beside each one write **THINGS I HAVE** tried, **THINGS I WILL** try and **THINGS I MIGHT** try. Don't be scared of trying something new.

And remember – "FAIL TO PREPARE AND PREPARE TO FAIL!"

Higher Chemistry

The Course

The main aims of the Higher Chemistry course are for learners to:

- develop and apply knowledge and understanding of chemistry
- develop an understanding of chemistry's role in scientific issues and relevant applications of chemistry, including the impact these could make in society and the environment
- develop scientific analytical thinking skills, including scientific evaluation, in a chemistry context
- develop the use of technology, equipment and materials, safely, in practical scientific activities, including using risk assessments
- develop scientific inquiry, investigative, problem solving and planning skills
- use and understand scientific literacy to communicate ideas and issues and to make scientifically informed choices
- develop skills of independent working.

How the Course is assessed

To gain the Course award:

(i) you must pass the four units – Chemical Changes and Structure, Researching Chemistry, Nature's Chemistry and Chemistry in Society. The units are assessed internally on a pass/fail basis.

(ii) you must submit an assignment which is externally marked by the SQA and is worth 20 marks.

(iii) you must sit the Higher Chemistry exam paper which is set and marked by the SQA and is worth 100 marks.

The course award is graded A–D, the grade being determined by the total mark you score in the examination and the mark you gain in the assignment.

The Examination

- The examination consists of one exam paper which has two sections and lasts 2 hours 30 minutes:

  Section 1 (multiple choice)   20 marks
  Section 2 (extended answer)   80 marks

Further details can be found in the Higher Chemistry section on the SQA website:
http://www.sqa.org.uk/sqa/47913.html

Key Tips For Your Success

Practise! Practise! Practise!

In common with Higher maths and the other Higher sciences, the key to exam success in Chemistry is to prepare by regularly answering questions. Use the questions as a prompt for further study: if you find that you cannot answer a question, review your notes and/or textbook to help you find the necessary knowledge to answer the question. You will quickly find out what you can/cannot do if you invest time attempting to answer questions. It is a much more valuable use of time than passively copying notes, which is a common trap many students fall into!

The data booklet

The data booklet contains formulae and useful data, which you will have to use in the exam. Although you might think that you have a good memory for chemical data (such as the symbols for elements or the atomic mass of an element) always check using the data booklet.

Calculations

In preparation for the exam, ensure that you recognise the different calculation types:

- relative rate
- using bond enthalpy
- using $cm\Delta T$
- percentage yield
- atom economy
- using molar volume
- volumetric calculations
- calculations from balanced chemical equations

You will encounter these calculations in the exam so it's worth spending time practising to ensure that you are familiar with the routines for solving these problems. Even if you are not sure how to attempt a calculation question, show your working! You will be given credit for calculations, which are relevant to the problem being solved.

Explain questions

You will encounter questions, which ask you to *explain your answer*. Take your time and attempt to explain to the examiner. If you can use a diagram or chemical equations to aid your answer, use these as they can really bring an answer to life.

Applying your knowledge of practical chemistry

As part of your Higher Chemistry experience, you should have had plenty of practice carrying out experiments using standard lab equipment and you should have had opportunities to evaluate your results from experiments. In the Higher exam, you are expected to be familiar with the techniques and apparatus listed in tables below.

Apparatus

Beaker	Dropper	Pipette filler
Boiling tube	Evaporating basin	Test tubes
Burette	Funnel	Thermometer
Conical flask	Measuring cylinder	Volumetric flask
Delivery tubes	Pipette	

Techniques

Distillation
Filtration
Methods for collecting a gas: over water or using a gas syringe
Safe heating methods: using a Bunsen, water bath or heating mantle
Titration
Use of a balance

The following general points about experimental chemistry are worth noting:

- A pipette is more accurate than a measuring cylinder for measuring fixed volumes of liquid. A burette can be used to measure non-standard volumes of liquid.
- A standard flask is used to make up a standard solution i.e. a solution of accurately known concentration. This is done by dissolving a known mass of solute in water and transferring to the standard flask with rinsings. Finally, the standard flask is made up to the mark with water.
- A gas syringe is an excellent method for measuring the volume of gas produced from an experiment.
- Bunsen burners cannot be used near flammable substances.
- A Bunsen burner does not allow you to control the rate of heating.

Analysis of data

From your experience working with experimental data you should know how to calculate averages, how to eliminate rogue data, how to draw graphs (scatter and best fit line/curve) and how to interpret graphs.

It is common in Higher exams to be presented with titration data such as the data shown in table below.

Titration	Volume of solution cm^3
1	26.0
2	24.1
3	39.0
4	24.2
5	24.8

Result 1 is a rough titration which is not accurate.

Results 2 and 4 could be used to calculate an average volume (= $24.15 cm^3$)

Result 3 is a rogue result and should be ignored.

Result 5 cannot be used to calculate the average volume as it is too far from 24.1 and 24.2 i.e. it is not accurate.

Numeracy

The Higher Chemistry exam will contain several questions that test your numeracy skills e.g. calculating relative rate, enthalpy changes, percentage yield etc. Other questions will ask you to "scale up" or "scale down" as this is a skill that is used by practising scientists in their day to day job.

Being able to deal with proportion is key to answering numeracy questions in chemistry. A common layout is shown in the examples below. In all cases, the unknown (what you are being asked to calculate) should be put on the right hand side.

[Example]

1.2g of methane burned to produce 52kJ of energy. Calculate the enthalpy of combustion of methane.

Answer:

This is really a proportion question. You have to understand that the enthalpy of combustion is the energy released when 1 mole of a substance is burned completely, and that 1 mole of methane is the gram formula mass i.e. 16g.

Step 1: State a relationship

Mass   Energy

1.2g ➡ 52kJ

(note that energy is placed on the right hand side as we want to calculate the energy)

Step 2: Scale to 1

1g ⇒ = 43.3kJ

Step 3: Calculate for the mass you are asked

16g ⇒ 16x 43.3= 693.3

Finally, answer the question: The ΔH combustion for methane = -693.3kJmol^{-1}

[Example]

A 100ml bottle of children's paracetamol costs £3.85. The ingredients label states that each 5ml dose contains 120mg of paracetamol. Calculate the cost per mg of paracetamol.

Answer:

Volume		Mass
5ml	⇒	120mg
1ml	⇒	24mg
100ml	⇒	2400mg

i.e. 1 bottle contains 2400mg of paracetamol

Mass		Cost
2400mg	⇒	£3.85
1mg	⇒	£0.0016

Open-ended questions

Real-life chemistry problems rarely have a fixed answer. In the Higher exam, you will encounter two 3 mark questions that are open-ended i.e. there is more than one "correct" answer. You will recognise these questions from the phrase *using your knowledge of chemistry* in the question. To tackle these, look at the following example.

[Example]

SQA Higher Chemistry 2013 Q.12

Cooking involves many chemical reactions. Proteins, fats, oils and esters are some examples of compounds found in food. A chemist suggested that cooking food could change compounds from being fat-soluble to water-soluble.

Use your knowledge of chemistry to comment on the accuracy of this statement.

Author's suggested answer

To tackle a question like this, focus on the key chemical words and think about the chemistry you know. What chemical reactions do you know that involve proteins, fats, oils and esters? Can you relate this to solubility?

Proteins - Long chain molecules linked by hydrogen bonding. Perhaps the proteins in food are insoluble as the chains are attracted to themselves. Cooking could cause the protein chains to untwist (breaking the hydrogen bonds) making them more likely to attract water to the exposed peptide links. In addition, cooking could cause the protein to hydrolyse to produce amino acids. Amino acids contain the polar amine group (–NH$_2$) and carboxyl group (–COOH) – both can form hydrogen bonds to water, therefore the amino acids can dissolve in water.

Fats and Oils - Insoluble in water as they are mainly large hydrocarbon structures. Fats and oils can hydrolyse to produce glycerol and fatty acids. Glycerol has three –OH groups so it could H-bond to water molecules and dissolve. Fatty acids contain a polar head (the carboxyl group-COOH) which is water soluble.

Esters - Non-polar and insoluble. Heating could hydrolyse the ester group producing an alcohol and carboxylic acid. Both of these molecules are polar and would dissolve in water.

A good answer for this question wouldn't have to contain all of the above. Indeed, it could focus on one molecule but give lots of detail. It's also a good idea to illustrate your answer with diagrams e.g. you could show typical structures and show how they can bond to water. If it enhances your answer by showing the examiner that you understand the chemistry, include it!

Good luck!

If you have followed the advice given in this introduction you will be well prepared for the Higher exam. When you sit the exam, take your time and use the experience as an opportunity to show the examiner how much you know. And good luck!

HIGHER FOR CfE

2014 Specimen Question Paper

SQ07/H/02

**Chemistry
Section 1 — Questions**

Date — Not applicable

Duration — 2 hours and 30 minutes

Reference may be made to the Chemistry Higher and Advanced Higher Data Booklet.

Instructions for the completion of Section 1 are given on *Page two* of your question and answer booklet SQ07/H/01.

Record your answers on the answer grid on *Page three* of your question and answer booklet.

Before leaving the examination room you must give your question and answer booklet to the Invigilator; if you do not you may lose all the marks for this paper.

SECTION 1 — 20 marks
Attempt ALL questions

1. Which type of bonding is **never** found in elements?

   A Metallic
   B London dispersion forces
   C Polar covalent
   D Non-polar covalent

2. In which of the following molecules will the chlorine atom carry a partial positive charge ($\delta+$)?

   A Cl–Br
   B Cl–Cl
   C Cl–F
   D Cl–I

3. Which of the following is **not** an example of a Van der Waals' force?

   A Covalent bonding
   B Hydrogen bonding
   C London dispersion forces
   D Permanent dipole-permanent dipole interactions

4. The diagram shows the melting points of successive elements across a period in the Periodic Table.

   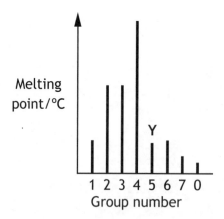

   Which of the following is a correct reason for the low melting point of element **Y**?

   A It has weak ionic bonds
   B It has weak covalent bonds
   C It has weakly-held outer electrons
   D It has weak forces between molecules

Page two

5. In which of the following will **both** changes result in an increase in the rate of a chemical reaction?

   A   A decrease in activation energy and an increase in the frequency of collisions

   B   An increase in activation energy and a decrease in particle size

   C   An increase in temperature and an increase in the particle size

   D   An increase in concentration and a decrease in the surface area of the reactant particles

6. Which of the following is **not** a correct statement about the effect of a catalyst?

   The catalyst

   A   provides energy so that more molecules have successful collisions

   B   lowers the energy that molecules need for successful collisions

   C   provides an alternative route to the products

   D   forms bonds with reacting molecules.

7. The graph shows how the rate of a reaction varies with the concentration of one of the reactants.

   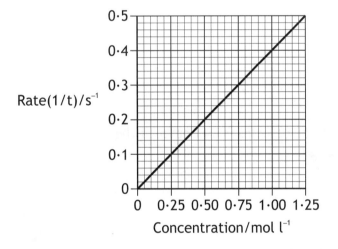

   Calculate the reaction time, in seconds, when the concentration of the reactant was 0.50 mol l^{-1}.

   A   0·2

   B   0·5

   C   2·0

   D   5·0

8. In which line of the table are fat, protein and soap correctly classified?

	Amides	Salts	Esters
A	Fat	Soap	Protein
B	Fat	Protein	Soap
C	Soap	Fat	Protein
D	Protein	Soap	Fat

9. The arrangement of amino acids in a peptide is Z-X-W-V-Y where the letters V, W, X, Y and Z represent amino acids.

   On partial hydrolysis of the peptide, which of the following sets of dipeptides is possible?

   A   V—Y, Z—X, W—Y, X—W
   B   Z—X, V—Y, W—V, X—W
   C   Z—X, X—V, W—V, V—Y
   D   X—W, X—Z, Z—W, Y—V

10.
$$H_3C - \underset{\underset{CH_3}{|}}{\overset{\overset{OH}{|}}{C}} - CH_3$$

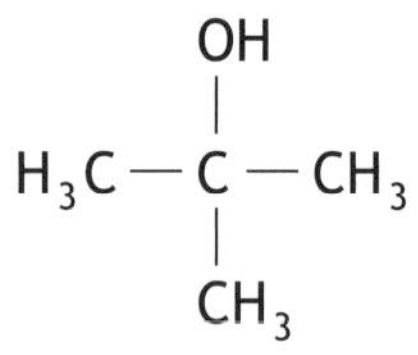

   Which of the following compounds is an isomer of the structure shown above?

   A   Butanal
   B   Butanone
   C   Butan-1-ol
   D   Butanoic acid

11. Erythrose can be used in the production of a chewing gum that helps prevent tooth decay.

$$HO-CH_2-\underset{\underset{OH}{|}}{CH}-\underset{\underset{OH}{|}}{CH}-C\overset{\nearrow O}{\searrow H}$$

Which of the following compounds will be the best solvent for erythrose?

A cyclohexane (H₂C-CH₂-CH₂-CH₂-CH₂-CH₂ ring)

B $CH_3-CH_2-CH_2-CH_2-CH_2-CH_3$

C CH_3-CH_2-OH

D $Cl-\underset{\underset{Cl}{|}}{\overset{\overset{H}{|}}{C}}-Cl$

12. Vanillin and zingerone are flavour molecules.

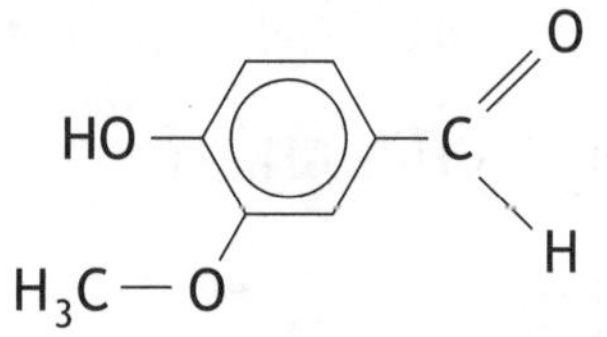

vanillin

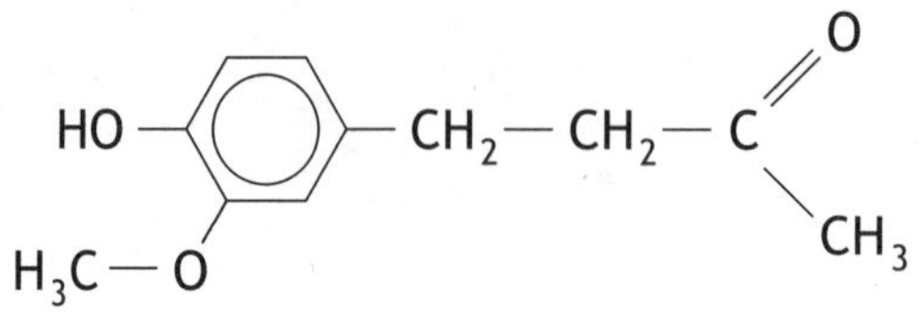

zingerone

Which line in the table correctly compares the properties of vanillin and zingerone?

	More soluble in water	More volatile
A	vanillin	vanillin
B	vanillin	zingerone
C	zingerone	vanillin
D	zingerone	zingerone

13. Soaps are produced by the following reaction.

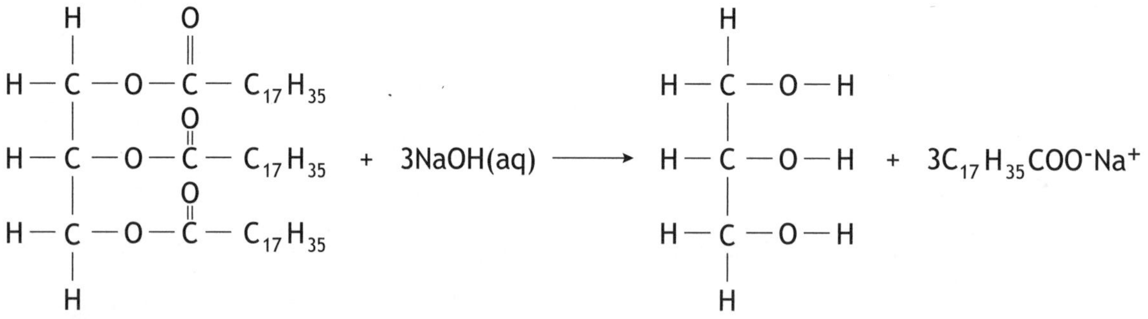

This reaction is an example of

A condensation
B esterification
C hydrolysis
D oxidation.

Page six

14. During a redox process in acid solution, iodate ions, $IO_3^-(aq)$, are converted into iodine, $I_2(aq)$.

$$IO_3^-(aq) \rightarrow I_2(aq)$$

The numbers of $H^+(aq)$ and $H_2O(\ell)$ required to balance the ion-electron equation for the formation of 1 mol of $I_2(aq)$ are, respectively

A 3 and 6

B 6 and 3

C 6 and 12

D 12 and 6.

15. $2SO_2(g) + O_2(g) \rightleftharpoons 2SO_3(g)$

The equation represents a mixture at equilibrium.

Which line in the table is true for the mixture after a further 2 hours of reaction?

	Rate of forward reaction	Rate of back reaction
A	decreases	decreases
B	increases	increases
C	unchanged	decreases
D	unchanged	unchanged

16. $5N_2O_4(\ell) + 4CH_3NHNH_2(\ell) \rightarrow 4CO_2(g) + 12H_2O(\ell) + 9N_2(g)$ $\Delta H = -5116\,kJ$

The energy released when 2 moles of each reactant are mixed and ignited is

A 1137 kJ

B 2046 kJ

C 2258 kJ

D 2843 kJ.

17. 1670 kJ of energy are given out when 2 moles of aluminium react completely with 1·5 moles of oxygen.

$$2Al(s) + 1\tfrac{1}{2}O_2(g) \rightarrow Al_2O_3(s)$$

The enthalpy of combustion of aluminium, in kJ mol^{-1}, is

- A −1113
- B −835
- C +835
- D +1113.

18. Which of the following elements is the strongest reducing agent?

- A Lithium
- B Bromine
- C Fluorine
- D Aluminium

19. 45 cm^3 of a solution could be most accurately measured out using a

- A 50 cm^3 beaker
- B 50 cm^3 burette
- C 50 cm^3 pipette
- D 50 cm^3 measuring cylinder.

20. Sulphur dioxide gas is denser than air and is very soluble in water.

Which of the following diagrams shows the most appropriate apparatus for collecting and measuring the volume of sulphur dioxide given off in a reaction?

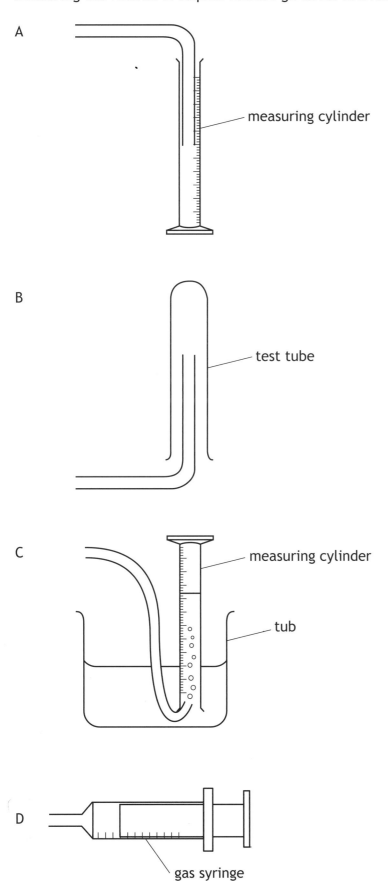

[END OF SECTION 1. NOW ATTEMPT THE QUESTIONS IN SECTION 2 OF YOUR QUESTION AND ANSWER BOOKLET.]

H

National Qualifications SPECIMEN ONLY

Mark

SQ07/H/01

**Chemistry
Section 1—Answer Grid
and Section 2**

Date — Not applicable

Duration — 2 hours and 30 minutes

Fill in these boxes and read what is printed below.

Full name of centre

Town

Forename(s)

Surname

Number of seat

Date of birth

Day   Month   Year

D D   M M   Y Y

Scottish candidate number

0 6 1 4 4 2 4 6 4

Reference may be made to the Chemistry Higher and Advanced Higher Data Booklet.

Total marks — 100

SECTION 1 — 20 marks

Attempt ALL questions.

Instructions for completion of Section 1 are given on *Page two*.

SECTION 2 — 80 marks

Attempt ALL questions

Write your answers clearly in the spaces provided in this booklet. Additional space for answers and rough work is provided at the end of this booklet. If you use this space you must clearly identify the question number you are attempting. Any rough work must be written in this booklet. You should score through your rough work when you have written your final copy.

Use **blue** or **black** ink.

Before leaving the examination room you must give this booklet to the Invigilator; if you do not you may lose all the marks for this paper.

SECTION 1—20 marks

The questions for Section 1 are contained in the question paper SQ07/H/02.
Read these and record your answers on the answer grid on *Page three* opposite.
DO NOT use gel pens.

1. The answer to each question is **either** A, B, C or D. Decide what your answer is, then fill in the appropriate bubble (see sample question below).

2. There is **only one correct** answer to each question.

3. Any rough working should be done on the additional space for answers and rough work at the end of this booklet.

Sample Question

To show that the ink in a ball-pen consists of a mixture of dyes, the method of separation would be:

    A   fractional distillation

    B   chromatography

    C   fractional crystallisation

    D   filtration.

The correct answer is **B**—chromatography. The answer **B** bubble has been clearly filled in (see below).

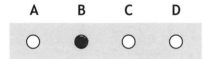

Changing an answer

If you decide to change your answer, cancel your first answer by putting a cross through it (see below) and fill in the answer you want. The answer below has been changed to **D**.

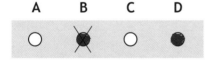

If you then decide to change back to an answer you have already scored out, put a tick (✓) to the **right** of the answer you want, as shown below:

SECTION 1 — Answer Grid

MARKS | DO NOT WRITE IN THIS MARGIN

SECTION 2 — 80 marks

Attempt ALL questions

1. Common salt, NaCl, is widely used in the food industry as a preservative and flavour enhancer.

(a) (i) Write the ion-electron equation for the first ionisation energy of sodium. **1**

(ii) **Explain clearly** why the first ionisation energy of sodium is much lower than its second ionisation energy. **3**

(b) The label on a tub of margarine states that 100 g of the margarine contains 0·70 g of sodium. The sodium is present as sodium chloride.

Calculate the mass of sodium chloride, in grams, present in a 10 g portion of the margarine.

The mass of one mole of sodium chloride, NaCl, is 58·5 g. **1**

2. (a) Nitrogen dioxide gas and carbon monoxide gas can react when molecules collide.

$$NO_2(g) + CO(g) \rightarrow NO(g) + CO_2(g)$$

State **two** conditions necessary for a collision to be successful.   2

(b) Hydrogen gas and chlorine gas react explosively in a photochemical reaction. In a demonstration experiment, the reaction was used to fire a table tennis ball across a room.

(i) A mixture of hydrogen gas and chlorine gas was generated by the electrolysis of hydrochloric acid.

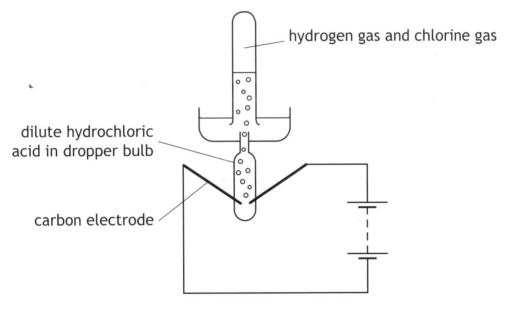

$$2HCl(aq) \rightarrow H_2(g) + Cl_2(g)$$

Calculate the number of moles of hydrochloric acid required to completely fill a 10 cm³ test tube with the hydrogen gas and chlorine gas mixture.

(Take the molar volume of a gas to be 24 litres mol⁻¹)   2

2. (b) (continued)

(ii) The filled test tube was fitted with a stopper to which a table tennis ball was attached. When a bright light was directed at the test tube, the gas mixture exploded and the ball was fired across the room.

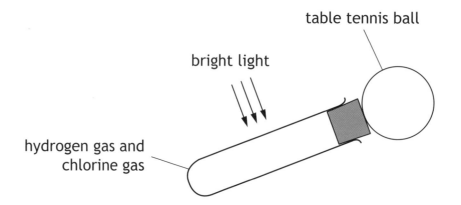

Chlorine reacts with hydrogen in a free radical chain reaction. Some steps in the chain reaction are shown in the table.

Reaction step	Name of step
$Cl_2 \rightarrow 2Cl\bullet$	
$Cl\bullet + H_2 \rightarrow HCl + H\bullet$ $H\bullet + Cl_2 \rightarrow HCl + Cl\bullet$	propagation
	termination

Complete the table by:

A inserting the missing name for the first step; **1**

B showing a possible termination step. **1**

2. (b) (continued)

(iii) The production of hydrogen chloride from hydrogen and chlorine is exothermic.

$$H_2(g) + Cl_2(g) \rightarrow 2HCl(g)$$

Using bond enthalpy values, calculate the enthalpy change, in kJ, for the reaction.

Show your working clearly. 2

3. A team of chemists is developing a fragrance for use in a shower gel for men.

(a) To give the gel a fruity smell the chemists are considering adding an ester.

They synthesise six isomeric esters. Volunteers smell each ester and give it a rating out of one hundred depending on how fruity the smell is.

Structure	Fruit-smell rating
$CH_3-C(=O)-O-CH_2-CH_2-CH_2-CH_2-CH_3$	100
$CH_3-C(=O)-O-CH(CH_3)-CH_2-CH_2-CH_3$	34
$CH_3-C(=O)-O-C(CH_3)_2-CH_2-CH_3$	0
$CH_3-CH_2-C(=O)-O-CH_2-CH_2-CH_2-CH_3$	92
$CH_3-CH(CH_3)-C(=O)-O-CH_2-CH_2-CH_3$	44
$(CH_3)_3C-C(=O)-O-CH_2-CH_3$	32

(i) Name the ester with the fruit-smell rating of 92.   **1**

3. (a) (continued)

(ii) Shown below are the structures of three more isomers.

Ester **A** $CH_3-CH_2-CH_2-C\begin{smallmatrix}\nearrow O\\ \searrow O-CH(CH_3)-CH_3\end{smallmatrix}$

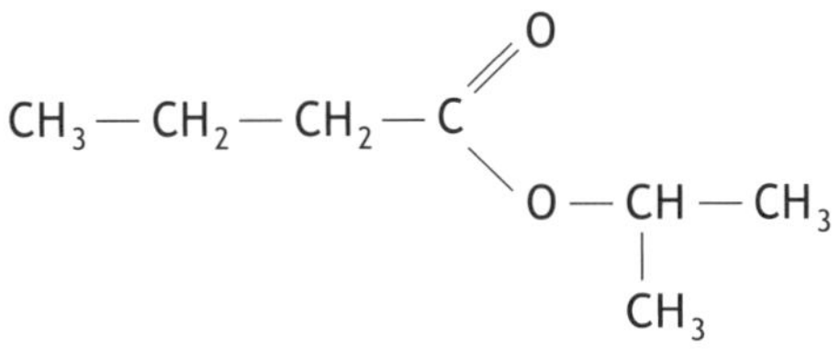

Ester **B** $CH_3-CH_2-CH_2-C\begin{smallmatrix}\nearrow O\\ \searrow O-CH_2-CH_2-CH_3\end{smallmatrix}$

Ester **C** $CH_3-CH_2-C\begin{smallmatrix}\nearrow O\\ \searrow O-C(CH_3)_3\end{smallmatrix}$

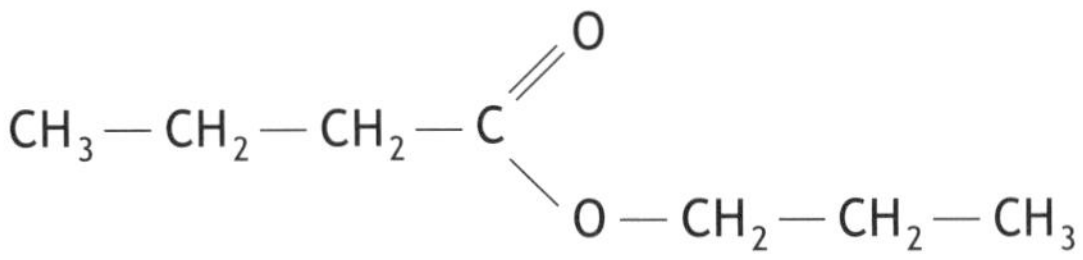

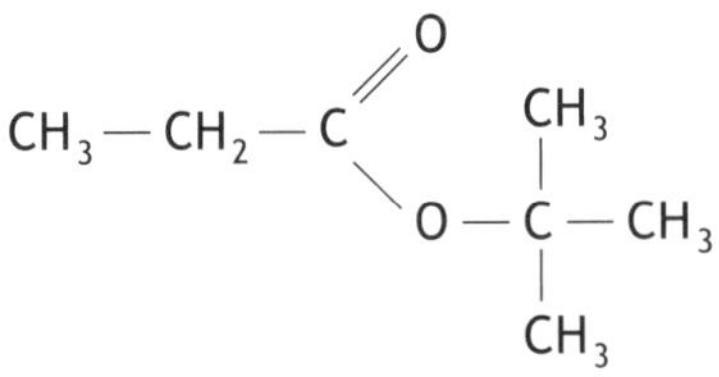

Arrange these esters in order of **decreasing** fruit-smell rating.

Ester ☐ > Ester ☐ > Ester ☐   1

(b) To create a fragrance for men, the compound civetone is added.

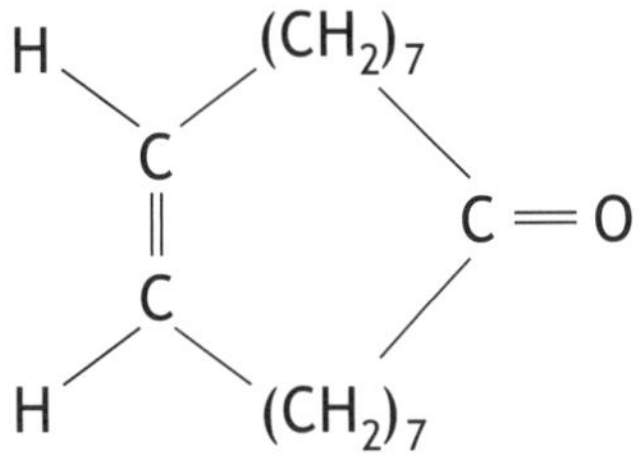

civetone (a cyclic compound with a C=C double bond between two CH groups, with (CH₂)₇ chains on each side connecting to a C=O group)

Draw a structural formula for the alcohol that can be oxidised to form civetone.   1

Page nine

3. (continued)

   (c) To make the shower gel produce a cold, tingling sensation when applied to the skin, menthol is added.

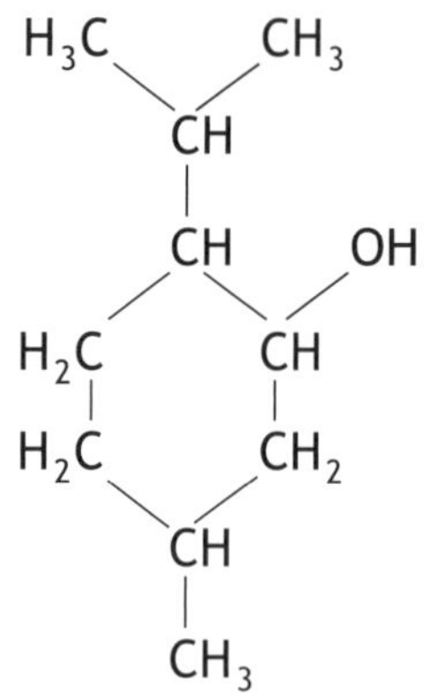

   [Structure of menthol showing a cyclohexane ring with an isopropyl group (CH(CH₃)₂), an OH group, and a CH₃ group]

   Like terpenes, menthol is formed from isoprene (2-methylbuta-1,3-diene).

   Circle an isoprene unit on the menthol structure above. **1**

4. Cooking changes the appearance and composition of foods.

   Using your knowledge of chemistry, comment on the changes to food that may occur during cooking. **3**

5. 2-Methylpropan-1-ol and ethanol are renewable fuels that are used as alternatives to petrol in car engines.

(a) A car was fuelled with 15 litres of ethanol. The ethanol burned to produce 351 000 kJ of energy.

Use the data in the table to calculate the volume of 2-methylpropan-1-ol that would burn to release the same energy.

Volume of 1 g of 2-methylpropan-1-ol	1·25 cm^3
Energy from 1 g of 2-methylpropan-1-ol	36·1 kJ

3

(b) Fuels containing alcohols have a tendency to absorb water, which can cause the engine to rust.

Water is absorbed by alcohols due to hydrogen bonds forming between the alcohol and water molecules.

In the box below, use a dotted line to show a hydrogen bond between a water molecule and 2-methylpropan-1-ol.

1

5. (continued)

(c) 2-Methylpropan-1-ol can also be converted to produce diesel and jet fuel.

The first step in the process is the production of 2-methylpropene.

$$C_4H_{10}O(\ell) \rightarrow C_4H_8(g) + H_2O(g)$$
2-methylpropan-1-ol   2-methylpropene

Using the data below, calculate the enthalpy change, in kJ mol^{-1}, for the production of 2-methylpropene from 2-methylpropan-1-ol.

$4C(s) + 5H_2(g) + \tfrac{1}{2}O_2(g) \rightarrow C_4H_{10}O(\ell)$ $\Delta H = -335$ kJ mol^{-1}

$4C(s) + 4H_2(g) \rightarrow C_4H_8(g)$ $\Delta H = -17$ kJ mol^{-1}

$H_2(g) + \tfrac{1}{2}O_2(g) \rightarrow H_2O(g)$ $\Delta H = -242$ kJ mol^{-1}

2

(d) If the viscosity of a fuel is not within a certain range then it can result in damage to the fuel pump and engine.

A student was asked to design an experiment to compare the viscosity of some fuels. Suggest an experiment that could be done to compare viscosities.

(You may wish to use a diagram to help with your description.)

2

6. Cyanoacrylate adhesives are a range of high performance "super glues".

   In its liquid form, super glue consists of cyanoacrylate monomers that rapidly polymerise in the presence of water to form a strong resin that joins two surfaces together.

   Cyanoacrylates have the general structure

   where R is a hydrocarbon group, eg -CH_3.

   (a) Some super glues contain methyl 2-cyanoacrylate.

   Circle the ester link in this structure. **1**

   (b) If used incorrectly, super glue can rapidly cause your fingers to stick together.

   (i) Suggest why super glue reacts rapidly on the surface of the skin. **1**

   (ii) Super glue can be removed from the skin using propanone as a solvent.

   Name the main type of van der Waals' forces that would be formed between propanone and super glue. **1**

6. (continued)

(c) Ethyl 2-cyanoacrylate is synthesised from ethyl 2-cyanoethanoate by a process based on the Knovenagel reaction.

$N\equiv C-CH_2-\overset{\overset{O}{\|}}{C}-O-CH_2CH_3$ + methanal → $N\equiv C-\overset{}{\underset{\underset{H\quad H}{\overset{\|}{C}}}{C}}-\overset{\overset{O}{\|}}{C}-O-CH_2CH_3$ + H_2O

| ethyl 2-cyanoethanoate mass of one mole = 113 g | reactant **A** mass of one mole = 30 g | ethyl 2-cyanoacrylate mass of one mole = 125 g | water mass of one mole = 18 g |

(i) Name reactant **A**.   1

(ii) Name this **type** of chemical reaction.   1

(iii) Calculate the atom economy for the formation of ethyl 2-cyanoacrylate using this process.

Show your working clearly.   2

6. (continued)

(d) The adhesive strength of super glue can be altered by introducing different alkyl groups to the monomer.

Hydrocarbon group	Shearing adhesive strength/N cm^{-2}
—CH$_3$	1800
—CH$_2$—CH$_3$	1560
—CH$_2$—CH$_2$—CH$_3$	930
—CH$_2$—CH$_2$—CH$_2$—CH$_3$	270
—CH(CH$_3$)—CH$_2$—CH$_3$ (with CH$_3$ branch)	420
—CH$_2$—CH=CH$_2$	1240
—CH$_2$—C≡CH	1670
—HC(CH$_3$)—C≡CH	1140

Estimate the adhesive strength of super glue that contains the monomer shown below.

1

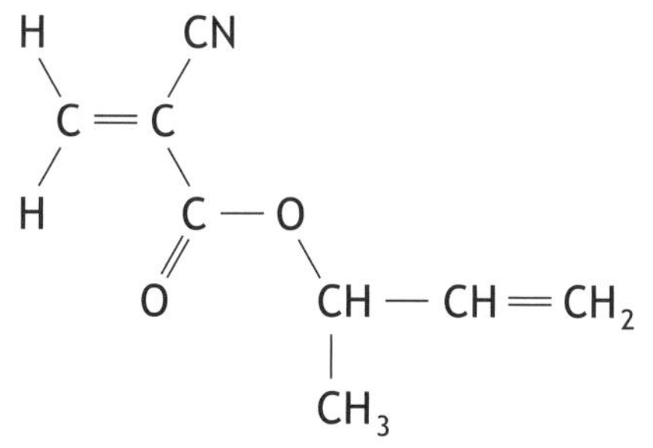

6. (continued)

   (e) Super glues have been developed for medical applications.

      (i) Medical tissue adhesive, containing octyl 2-cyanoacrylate, can be used for wound closures instead of sutures or stitches.

      Draw a structural formula for octyl 2-cyanoacrylate. **1**

      (ii) The graph below compares the temperature change during the polymerisation reaction for two different brands of medical tissue adhesive.

      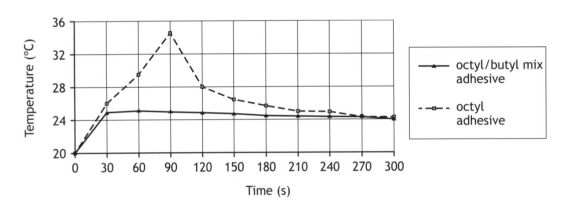

      Suggest an advantage to the patient of using the octyl/butyl mix adhesive. **1**

7. A student analysed a local water supply to determine fluoride and nitrite ion levels.

   (a) The concentration of fluoride ions in water was determined by adding a red coloured compound that absorbs light to the water samples. The fluoride ions reacted with the red compound to produce a colourless compound. Higher concentrations of fluoride ions produce less coloured solutions which absorb less light.

   The student initially prepared a standard solution of sodium fluoride with fluoride ion concentration of 100 mg l^{-1}.

   (i) State what is meant by the term **standard solution**. **1**

   (ii) Describe how the standard solution is prepared from a weighed sample of sodium fluoride. **2**

   (iii) Explain why the student should use distilled or deionised water rather than tap water when preparing the standard solution. **2**

7. (a) (continued)

(iv) The student prepared a series of standard solutions of fluoride ions and reacted each with a sample of the red compound. The light absorbance of each solution was measured and the results graphed.

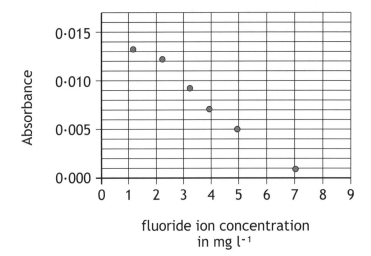

fluoride ion concentration in mg l⁻¹

Determine the concentration of fluoride ions in a solution with absorbance 0·012.

7. (continued)

(b) The concentration of nitrite ions in the water supply was determined by titrating water samples with acidified permanganate solutions.

(i) An average of 21·6 cm³ of 0·015 mol l⁻¹ acidified permanganate solution was required to react completely with the nitrite ions in a 25·0 cm³ sample of water.

The equation for the reaction taking place is

$$2MnO_4^-(aq) + 5NO_2^-(aq) + 6H^+(aq) \rightarrow 2Mn^{2+}(aq) + 5NO_3^-(aq) + 3H_2O(\ell)$$

Calculate the nitrite ion concentration, in mol l⁻¹, in the water.

Show your working clearly. 3

(ii) During the reaction the nitrite ion is oxidised to the nitrate ion. Complete the ion-electron equation for the oxidation of the nitrite ion.

$NO_2^-(aq) \rightarrow NO_3^-(aq)$ 1

MARKS | DO NOT WRITE IN THIS MARGIN

8. Ibuprofen is one of the best-selling pain killers in the UK.

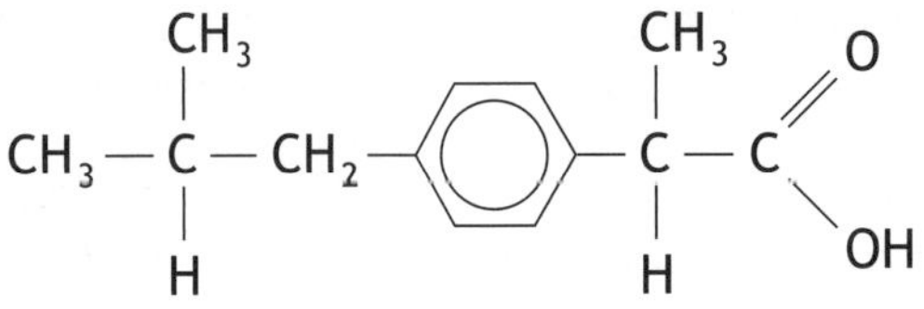

(a) Ibuprofen tablets should not be taken by people who suffer from acid indigestion. Name the functional group present in ibuprofen that makes this drug unsuitable for these people. **1**

(b) Ibuprofen is normally taken as tablets or pills and it is only slightly soluble in water.

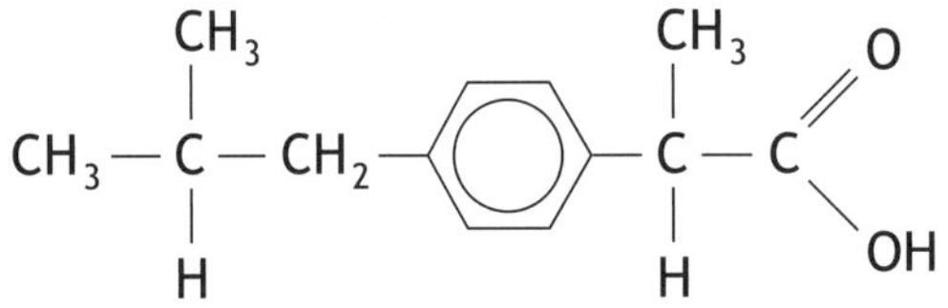

(i) Suggest why ibuprofen is only slightly soluble in water. **1**

(ii) Ibuprofen is also available as an "infant formula" emulsion for young children.

The emulsifier used is polysorbate 80. Its structure is shown below.

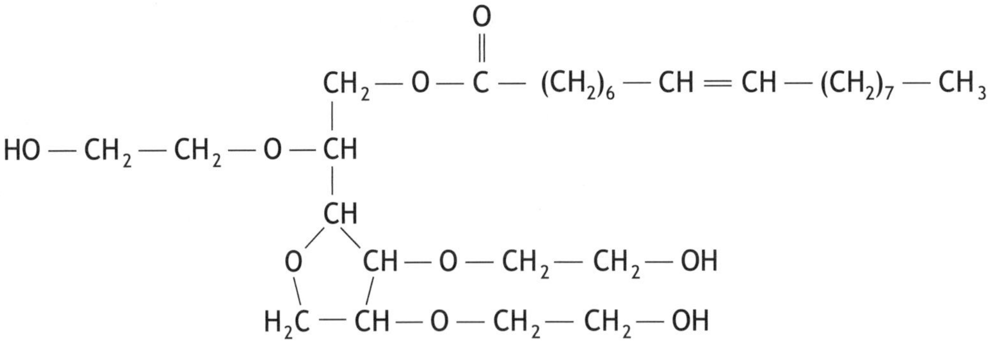

Circle the part of the polysorbate 80 molecule that is hydrophobic. **1**

8. (b) (continued)

(iii) The emulsion contains 2·0 g of ibuprofen in every 100 cm³ of emulsion.

The recommended dose for treating a three month old baby is 0·050 g.

Calculate the volume, in cm³, of "infant formula" needed to treat a three month old baby. **1**

(c) Paracetamol is another widely used painkiller. Its structure is shown below.

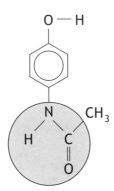

(i) Name the functional group shaded in the structure. **1**

8. (c) (continued)

(ii) The concentration of paracetamol in a solution can be determined by measuring how much UV radiation it absorbs. The quantity of UV radiation absorbed is directly proportional to the concentration of paracetamol.

The graph shows how the absorbance of a sample containing $0.040\,g\,l^{-1}$ paracetamol varies with wavelength.

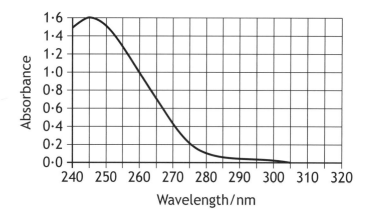

The absorbance of a second sample of paracetamol solution measured at 245 nm was 0·90.

Determine the concentration, in $g\,l^{-1}$, of this second paracetamol solution.

9. A student carried out some experiments using different fats and oils.

(a) The first experiment allowed the iodine number to be calculated. The iodine number is the mass of iodine, in grams, that will react with 100 g of the fat or oil. The student's results are shown.

Fat or oil	Iodine number	Typical molecule found in the fat or oil
Olive oil	84	$H_{33}C_{17}-C(=O)-O-CH$ with $H_2C-O-C(=O)-C_{17}H_{33}$ and $H_2C-O-C(=O)-C_{17}H_{33}$
Shea butter	43	$H_{33}C_{17}-C(=O)-O-CH$ with $H_2C-O-C(=O)-C_{17}H_{35}$ and $H_2C-O-C(=O)-C_{17}H_{33}$
Linseed oil	172	$H_{29}C_{17}-C(=O)-O-CH$ with $H_2C-O-C(=O)-C_{17}H_{33}$ and $H_2C-O-C(=O)-C_{17}H_{31}$
Sunflower oil		$H_{31}C_{17}-C(=O)-O-CH$ with $H_2C-O-C(=O)-C_{17}H_{33}$ and $H_2C-O-C(=O)-C_{17}H_{31}$

(i) Shea butter is a solid at room temperature.

Explain why the melting point of shea butter is higher than room temperature.

2

9. (a) (continued)

   (ii) Predict the iodine number of sunflower oil. **1**

   (iii) Name the substance that reacts with oils to turn them rancid. **1**

(b) In the second experiment some oils were used to make soap. The oil, triolein, was reacted with sodium hydroxide.

$$(C_{17}H_{33}COO)_3C_3H_5 \;+\; 3NaOH \;\rightarrow\; 3C_{17}H_{33}COONa \;+\; X$$

triolein sodium oleate
mass of one mole = 884 g mass of one mole = 304 g

   (i) Name product X. **1**

   (ii) 5·0 g of triolein was dissolved in ethanol and placed in a test tube with excess sodium hydroxide. The mixture was heated to 80 °C.

   State a suitable method for heating the reaction mixture. **1**

   (iii) The experiment produced 1·28 g of sodium oleate.

   Calculate the percentage yield. **2**

10. Urea, $(NH_2)_2CO$, is an important industrial chemical that is mainly used in fertilisers. It is made by the Bosch-Meiser process.

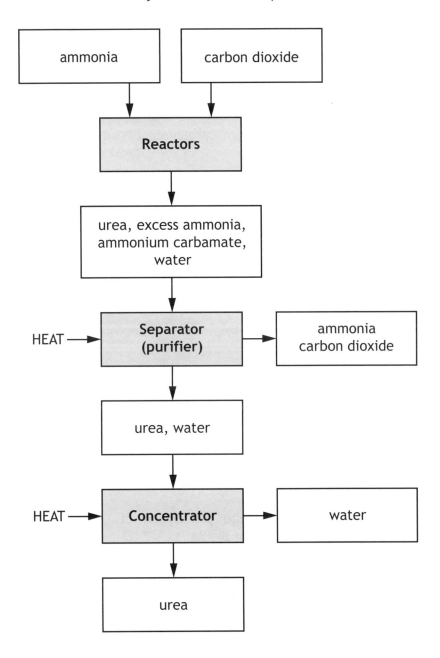

10. (continued)

(a) (i) In the reactors, the production of urea involves two reversible reactions.

In the first reaction ammonium carbamate is produced.

$$2NH_3(g) + CO_2(g) \rightleftharpoons H_2NCOONH_4(g)$$

In the second reaction the ammonium carbamate decomposes to form urea.

$$H_2NCOONH_4(g) \rightleftharpoons (NH_2)_2CO(g) + H_2O(g)$$

A chemical plant produces 530 tonnes of urea per day.

Calculate the theoretical mass, in tonnes, of ammonia required to produce 530 tonnes of urea. **2**

(ii) An undesirable side reaction is the production of biuret, a compound that can burn the leaves of plants.

$$2(NH_2)_2CO(aq) \rightleftharpoons \underset{\text{biuret}}{NH_2CONHCONH_2(aq)} + NH_3(g)$$

State why having an excess of ammonia in the reactors will decrease the amount of biuret produced. **1**

10. (a) (continued)

(iii) In the separator the ammonium carbamate from the reactors decomposes to form ammonia and carbon dioxide.

$$NH_2COONH_4(aq) \rightleftharpoons 2NH_3(g) + CO_2(g)$$

Explain clearly why a low pressure is used in the separator. **2**

(iv) Add a line to the flow chart, to show how the Bosch-Meiser process can be made more economical. **1**

(b) Soil bacteria are mainly responsible for releasing nitrogen in urea into the soil so that it can be taken up by plants. The first stage in the process is the hydrolysis of urea using the enzyme urease.

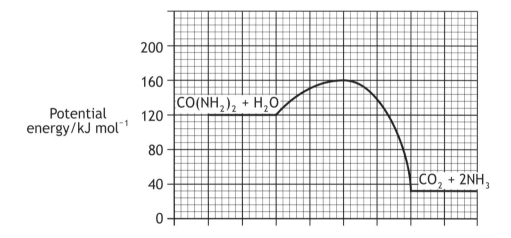

(i) Determine the enthalpy change, in kJ mol⁻¹, for the reaction. **1**

(ii) Acid is a less effective catalyst than urease for this reaction. Add a curve to the potential energy diagram to show the change in potential energy when acid is used as the catalyst. **1**

MARKS DO NOT WRITE IN THIS MARGIN

11. A TV programme was reproducing a pharmacy from the 19th century and planned to use the original 19th century pharmacy jars that had been kept in a museum. The TV company wanted to know what compounds the jars were likely to contain now.

Substances used in pharmacies over a hundred years ago included:

- Essential oils dissolved in ethanol.

  Some molecules included in these essential oils were:

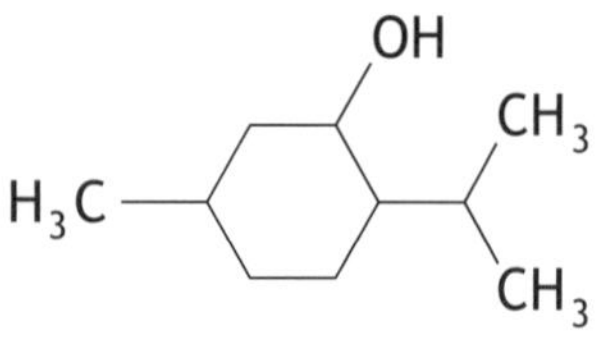

menthol

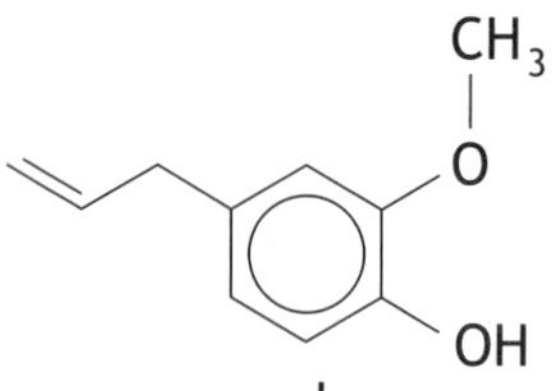

eugenol

- Aspirin.

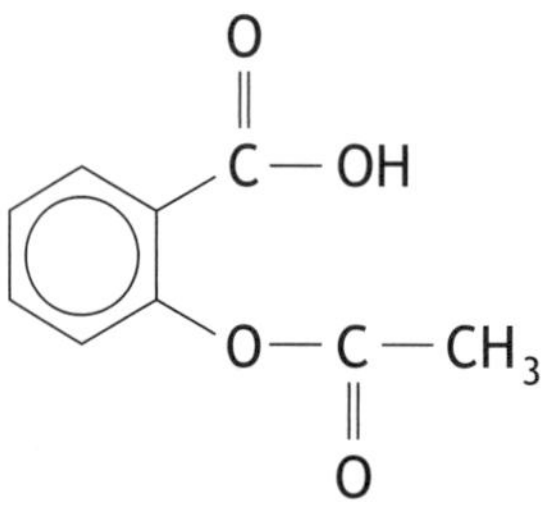

- Ointments that contained animal fats like lard, beef fat or beeswax.

Using your knowledge of chemistry, comment on what compounds the old pharmacy jars might contain now. 3

12. Proteins are an important part of a healthy diet because they provide essential amino acids.

(a) State what is meant by an **essential amino acid**. [1]

(b) Eggs and fish are good dietary sources of the essential amino acid, methionine.

The recommended daily allowance of methionine for an adult is 15 mg per kg of body mass.

Tuna contains 755 mg of methionine per 100 g portion.

Calculate the mass, in grams, of tuna that would provide the recommended daily allowance of methionine for a 60 kg adult. [2]

12. (continued)

(c) Mixtures of amino acids can be separated using paper chromatography. On a chromatogram, the retention factor, R_f, for a substance can be a useful method of identifying the substance.

$$R_f = \frac{\text{distance moved by the substance}}{\text{maximum distance moved by the solvent}}$$

(i) A solution containing a mixture of four amino acids was applied to a piece of chromatography paper that was then placed in solvent 1.

Chromatogram 1 is shown below.

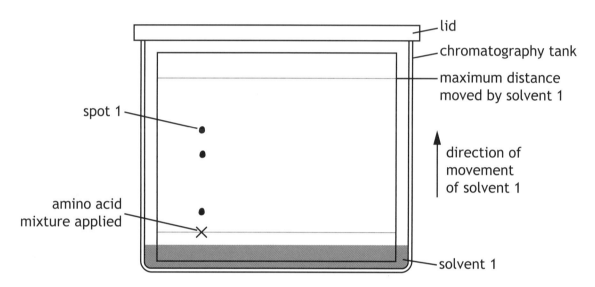

Amino Acid	R_f (solvent 1)
alanine	0·51
arganine	0·16
threonine	0·51
tyrosine	0·68

Identify the amino acid that corresponds to spot 1 on the chromatogram.

12. (c) (continued)

(ii) The chromatogram was dried, rotated through 90° and then placed in solvent 2.

Chromatogram 2 is shown below.

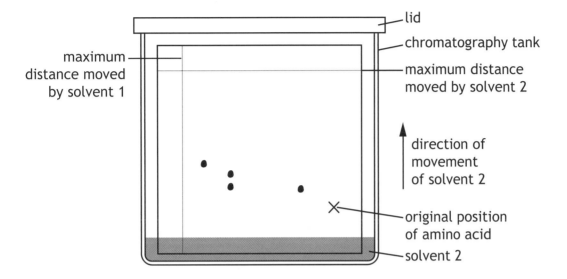

Amino Acid	R_f (solvent 2)
alanine	0·21
arganine	0·21
threonine	0·34
tyrosine	0·43

The retention factors for each of the amino acids in solvent 2 are shown in the table.

Draw a circle around the spot on chromatogram 2 that corresponds to the amino acid alanine. **1**

(iii) Explain why only three spots are present in chromatogram 1 while four spots are present in chromatogram 2. **2**

[END OF SPECIMEN QUESTION PAPER]

ADDITIONAL SPACE FOR ANSWERS AND ROUGH WORK

ADDITIONAL SPACE FOR ANSWERS AND ROUGH WORK

HIGHER FOR CfE

Model Paper 1

Whilst this Model Paper has been specially commissioned by Hodder Gibson for use as practice for the Higher (for Curriculum for Excellence) exams, the key reference documents remain the SQA Specimen Paper 2014 and SQA Past Paper 2015.

Chemistry
Section 1 — Questions

Duration — 2 hours and 30 minutes

Reference may be made to the Chemistry Higher and Advanced Higher Data Booklet.

Instructions for the completion of Section 1 are given on *Page two* of your question and answer booklet.

Record your answers on the answer grid on *Page three* of your question and answer booklet.

Before leaving the examination room you must give your question and answer booklet to the Invigilator; if you do not you may lose all the marks for this paper.

SECTION 1 — 20 marks
Attempt ALL questions

1. Which of the following would lead to a decrease in reaction rate for the reaction between hexene and hydrogen gas?

   A   Increasing the pressure of hydrogen gas.

   B   Increasing the activation energy of the reaction.

   C   Increasing the temperature.

   D   Improving the collision geometry.

2. The graph shows the energy distribution of molecules at two different temperatures, T_1 and T_2.

   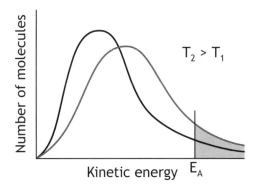

   Which statement is the best description for the increase in reaction rate as the temperature is increased from T_1 to T_2?

   A   The activation energy decreased as the temperature increased.

   B   The collision geometry improved as the temperature increased.

   C   More molecules have the minimum energy required to react.

   D   The molecules are moving at a faster rate.

3. An element with a melting point >1000°C, does not conduct electricity as a solid and forms a compound with oxygen which has a melting point >1000°C.

   The element is

   A   carbon

   B   iron

   C   silicon

   D   phosphorus.

4. The covalent radius of potassium is greater than the covalent radius of lithium because

   A potassium has a higher nuclear charge
   B there is a greater force of attraction between the potassium nucleus and its electrons
   C the screening of nuclear charge is greater in potassium than it is in lithium
   D lithium is more reactive than potassium.

5. Which of the following compounds has most covalent character?

   A PF_3
   B PH_3
   C NF_3
   D NH_3

6. Which of the following represents the second ionisation energy of chlorine?

   A $Cl_2(g) \rightarrow Cl_2^{2+}(g) + 2e^-$
   B $Cl(g) \rightarrow Cl g^{2+}(g) + 2e^-$
   C $Cl(g) \rightarrow Cl g^{2+}(g) + e^-$
   D $Cl(g)^+ \rightarrow Cl g^{2+}(g) + e^-$

7. Which of the following bonding types does not have a difference in electronegativity?

   A Metallic
   B Polar covalent
   C Dipole-dipole
   D Ionic

8. Which of the following statements **cannot** be applied to glycerol?

   A It is also known as propane-1,2,3-triol.
   B It is formed by hydrating fats or oils.
   C It has a high boiling point because of the hydrogen bonding between molecules.
   D It can react with fatty acids to form fats and oils.

9. Which three functional groups are present in the compound shown below?

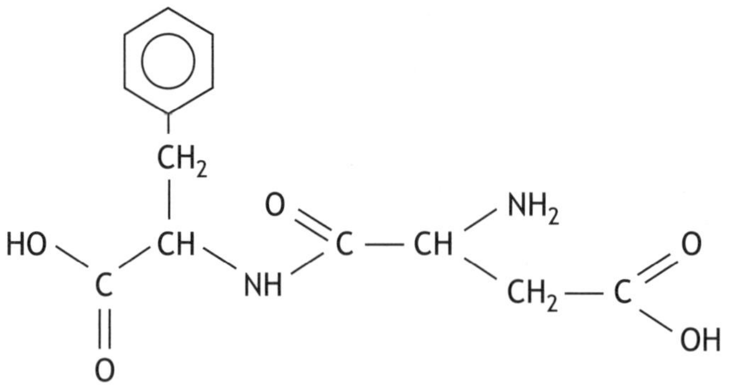

   A carboxyl, amine and alcohol
   B amide, carboxyl and amine
   C aldehyde, peptide and amine
   D carboxyl, peptide and alcohol

10. The structure shown below is an example of

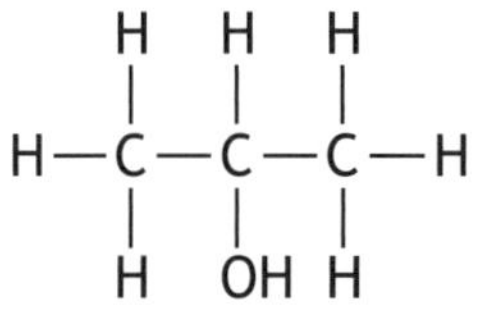

    A a primary alcohol
    B a ketone
    C a secondary alcohol
    D an aldehyde.

11. Denaturing a protein involves

    A hydrolysing the peptide bonds
    B joining amino acids to form a new protein
    C breaking covalent bonds
    D breaking hydrogen bonds.

12. Limonene is a terpene that is present in oranges and lemons.

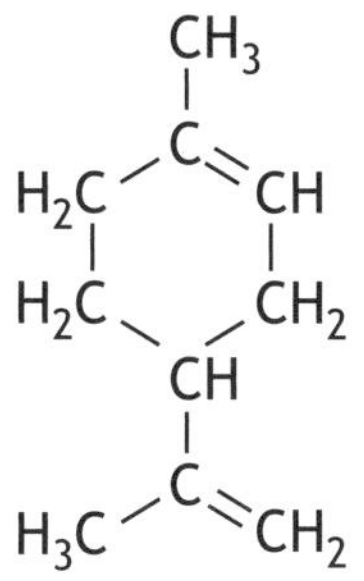

$$\begin{array}{c} CH_3 \\ | \\ H_2C-C=CH \\ | \quad\quad | \\ H_2C\quad\quad CH_2 \\ \diagdown CH \diagup \\ | \\ H_3C-C=CH_2 \end{array}$$

The number of isoprene units found in limonene is

A 1

B 2

C 3

D 4.

13. A free radical scavenger can

A initiate a free radical reaction

B terminate a free radical reaction

C increase the number of free radicals in a molecule

D form unstable molecules by reacting with free radicals.

14. Which of the following reactions has an atom economy of 100%?

A $HCl(aq) + NaOH(aq) \rightarrow NaCl(aq) + H_2O(l)$

B $Na_2O(s) + H_2SO_4(aq) \rightarrow Na_2SO_4(aq) + H_2O(l)$

C $CaCO_3(s) \rightarrow CaO(s) + CO_2(g)$

D $C_2H_4(g) + H_2O(l) \rightarrow C_2H_5OH(l)$

15. $CH_4(g) + 2O_2(g) \rightarrow CO_2(g) + 2H_2O(l)$

The volume of gas produced from the complete combustion of 200cm³ of methane in excess oxygen, at room temperature, is

A 200cm³

B 300cm³

C 600cm³

D 100cm³.

16. $2SO_2(g) + O_2(g) \rightleftharpoons 2SO_3(g)$

The formation of sulfur trioxide from sulfur dioxide and oxygen is an exothermic reaction.

Which of the following conditions would lead to the equilibrium shifting to the right?

- A high temperature and high pressure
- B high temperature and low pressure
- C low temperature and low pressure
- D low temperature and high pressure

17. Which of the following shows the correct equation for the enthalpy of combustion of methane?

- A $CH_4(g) + 1\frac{1}{2}O_2(g) \rightarrow CO(g) + 2H_2O(l)$
- B $2CH_4(g) + 3O_2(g) \rightarrow 2CO(g) + 4H_2O(l)$
- C $CH_4(g) + 2O_2(g) \rightarrow CO_2(g) + 2H_2O(l)$
- D $4CH_4(g) + 8O_2(g) \rightarrow 4CO_2(g) + 8H_2O(l)$

18. Which of the following substances is the strongest oxidising agent?

- A Br_2
- B Cl_2
- C I_2
- D F_2

19. Iodate ions can be converted into iodine: $IO_3^-(aq) \rightarrow I_2(aq)$

The number of $H^+(aq)$ and $H_2O(l)$ required to balance this ion electron equation for the formation of 1 mol of $I_2(aq)$ are

- A 12 and 6
- B 6 and 3
- C 3 and 6
- D 6 and 12

20. A standard solution is a solution with an accurately known

- A volume
- B mass
- C concentration
- D temperature.

[END OF SECTION 1. NOW ATTEMPT THE QUESTIONS IN SECTION 2 OF YOUR QUESTION AND ANSWER BOOKLET.]

National Qualifications
MODEL PAPER 1

Chemistry
Section 1—Answer Grid and Section 2

Duration — 2 hours and 30 minutes

Reference may be made to the Chemistry Higher and Advanced Higher Data Booklet.

Total marks — 100

SECTION 1 — 20 marks

Attempt ALL questions.

Instructions for completion of Section 1 are given on *Page 64*.

SECTION 2 — 80 marks

Attempt ALL questions

Write your answers clearly in the spaces provided in this booklet. Additional space for answers and rough work is provided at the end of this booklet. If you use this space you must clearly identify the question number you are attempting. Any rough work must be written in this booklet. You should score through your rough work when you have written your final copy.

Use **blue** or **black** ink.

Before leaving the examination room you must give this booklet to the Invigilator; if you do not you may lose all the marks for this paper.

SECTION 1—20 marks

The questions for Section 1 are contained on *Page 57* — Questions.
Read these and record your answers on the answer grid on *Page 65* opposite.
DO NOT use gel pens.

1. The answer to each question is **either** A, B, C or D. Decide what your answer is, then fill in the appropriate bubble (see sample question below).

2. There is **only one correct** answer to each question.

3. Any rough working should be done on the additional space for answers and rough work at the end of this booklet.

Sample Question

To show that the ink in a ball-pen consists of a mixture of dyes, the method of separation would be:

    A   fractional distillation

    B   chromatography

    C   fractional crystallisation

    D   filtration.

The correct answer is **B**—chromatography. The answer **B** bubble has been clearly filled in (see below).

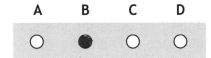

Changing an answer

If you decide to change your answer, cancel your first answer by putting a cross through it (see below) and fill in the answer you want. The answer below has been changed to **D**.

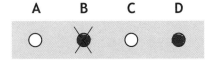

If you then decide to change back to an answer you have already scored out, put a tick (✓) to the **right** of the answer you want, as shown below:

SECTION 1 — Answer Grid

SECTION 2 — 80 marks

Attempt ALL questions

1. The elements of the Periodic Table can be categorised according to their bonding and structure.

    (a) Complete the table below for (i) — (iv) **2**

Bonding and Structure at room temperature	Example of an Element with this bonding and structure
Metallic	(i)
(ii)	Silicon
Covalent molecular solid	(iii)
(iv)	Neon

    (b) Compounds of Lithium are commonly used as medications.

    (i) Write an ion-electron equation for the first ionisation energy of lithium. **1**

    (ii) Explain clearly why the second ionisation energy of Lithium is much higher than the first ionisation energy of lithium. **3**

    (iii) A prescription for a medication stated that each tablet contained 400mg of lithium carbonate, Li_2CO_3.

    Calculate the mass, in mg, of lithium present in 400mg of lithium carbonate.

    The mass of one mole of Li_2CO_3 is 73.8g **1**

1. (continued)

   (c) The Periodic Table groups together elements with similar properties. In most Periodic Tables hydrogen is placed at the top of Group 1, but in some it is placed at the top of Group 7.

   Using your knowledge of Chemistry, comment on why hydrogen can be placed at the top of Group 1 and Group 7. **3**

2. Swimming pools can contain a variety of compounds such as nitrogen trichloride and ammonia.

   Nitrogen trichloride and ammonia have similar structures, but different properties.

   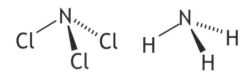

   Ammonia is more soluble in water than nitrogen trichloride.

   (a) Explain clearly why ammonia is more soluble in water than nitrogen trichloride.

   Your answer should include the names of the intermolecular forces involved. **3**

   (b) When it reacts with water, nitrogen trichloride forms ammonia and hypochlorus acid.

   $NCl_3(g) + H_2O(l) \rightarrow NH_3(aq) + HOCl(aq)$

   Balance the equation for this reaction. **1**

2. (continued)

   (c) In the presence of light, hypochlorous acid decomposes by a free-radical chain reaction to produce hydrochloric acid and oxygen.

   2HOCl(aq) → 2HCl(aq) + O$_2$(g)

   (i) What is meant by the term *free radical*? **1**

   (ii) The equation shows one of the steps in the free radical chain reaction.

   HOCl → HO• + Cl•

   What term describes this type of step in the free radical chain reaction? **1**

   (iii) The rate of decomposition of hypochlorous acid is increased if copper oxide is added.

   The diagram below shows the change in potential energy during this reaction when carried out without copper oxide.

   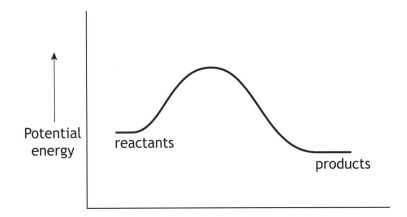

   Add a line to the diagram to show the effect of adding copper oxide to the reaction. **1**

   (d) Ammonia can be prepared by reacting nitrogen with hydrogen.

   The equation for this reaction is shown.

   N$_2$(g) + 3H$_2$(g) → 2NH$_3$(g)

   Using bond enthalpy values from the data booklet, calculate the enthalpy change for this reaction. **2**

2. (continued)

(e) For a reaction between nitrogen and hydrogen, the two molecules must collide. State one other condition necessary for a successful reaction. **1**

Total marks **10**

3. (a) In some countries, ethanol is used as a substitute for petrol. This ethanol is produced by fermentation of glucose, using yeast enzymes.

During the fermentation process, glucose is first converted into pyruvate. The pyruvate is then converted to ethanol in a two-step process.

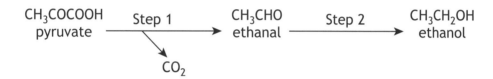

(i) **Step 1** is catalysed by an enzyme. Enzymes are proteins that can act as catalysts because they have a specific shape.

Why, when the temperature is raised above a certain value, does the rate of reaction decrease? **1**

(ii) Why can **Step 2** be described as a reduction reaction? **1**

MARKS | DO NOT WRITE IN THIS MARGIN

3. (a) (continued)

(iii) The overall equation for the fermentation of glucose is

$C_6H_{12}O_6$	$\rightarrow$	$2C_2H_5OH$	+	$2CO_2$
mass of one mole = 180 g		mass of one mole = 46 g		

Calculate the percentage yield of ethanol if 445 g of ethanol is produced from 1·0 kg of glucose. 3

Show your working clearly.

(b) The energy density value of a fuel is the energy released when one kilogram of the fuel is burned.

The enthalpy of combustion of ethanol is −1367 kJ mol^{-1}.

Calculate the energy density value, in kJ kg^{-1}, of ethanol. 1

3. (continued)

(c) The quantity of alcohol present after a fermentation reaction is called the % alcohol by volume.

This can be calculated from measurements taken using an instrument called a hydrometer. The hydrometer is floated in the liquid sample, before and after fermentation, to measure its specific gravity.

% alcohol by volume = change in specific gravity of liquid x f

where f is a conversion factor, which varies as shown in the table.

Change in specific gravity of liquid	f
Up to 6·9	0·125
7·0 — 10·4	0·126
10·5 — 17·2	0·127
17·3 — 26·1	0·128
26·2 — 36·0	0·129
36·1 — 46·5	0·130
46·6 — 57·1	0·131

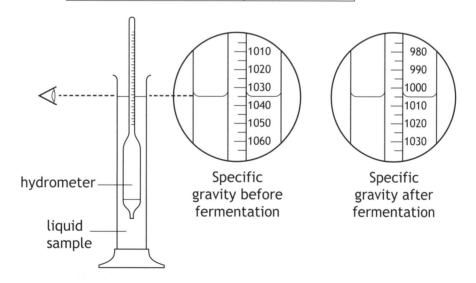

Calculate the % alcohol by volume for this sample.

Total marks 8

4. Limonene and geraniol are examples of terpenes found in essential oils.

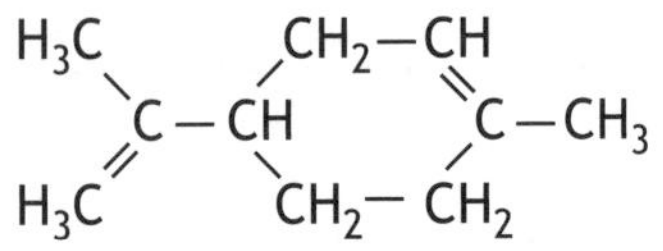

$C_{10}H_{16}$
limonene

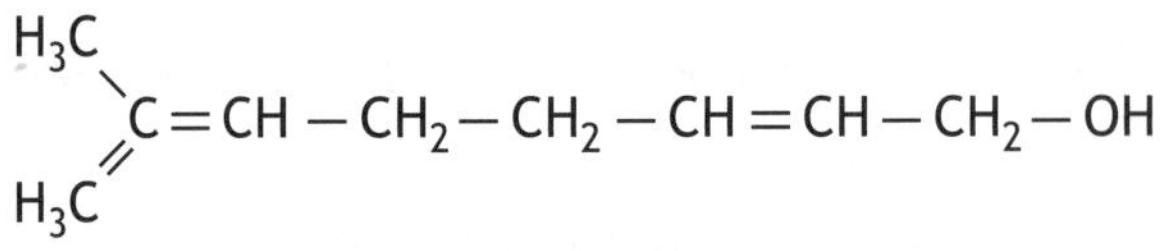

$C_9H_{16}O$
geraniol

(a) Draw the structure of isoprene. **1**

(b) Why does geraniol evaporate more slowly than limonene? **1**

(c) The structure of one of the first synthetic scents used in perfume is shown below.

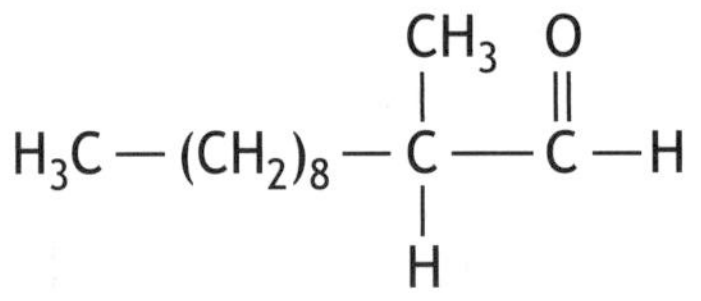

$$H_3C-(CH_2)_8-\underset{\underset{H}{|}}{\overset{\overset{CH_3}{|}}{C}}-\overset{\overset{O}{\|}}{C}-H$$

(i) Name the family of carbonyl compounds to which this synthetic scent belongs. **1**

(ii) Complete the structure below to show the product formed when this scent is oxidised. **1**

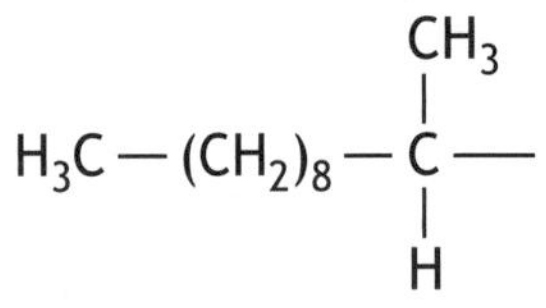

$$H_3C-(CH_2)_8-\underset{\underset{H}{|}}{\overset{\overset{CH_3}{|}}{C}}-$$

Total marks **4**

MARKS | DO NOT WRITE IN THIS MARGIN

5. The label shown below lists some of the ingredients found in an ice-cream.

Emulsifier (Mono- and Di-Glycerides of Fatty acids), Thickener (Carboxymethylcellulose), Flavouring (pentyl butanoate), Antioxidant (Ascorbic acid), Milk proteins

A student suggested that the compounds listed were used to enhance the flavour, solubility and shelf-life of the ice-cream. **Using your knowledge of chemistry**, comment on the accuracy of this statement. **3**

6. Soft drinks contain many ingredients.

(a) Aspartame is added to many soft drinks as a sweetener. Its structure is shown below.

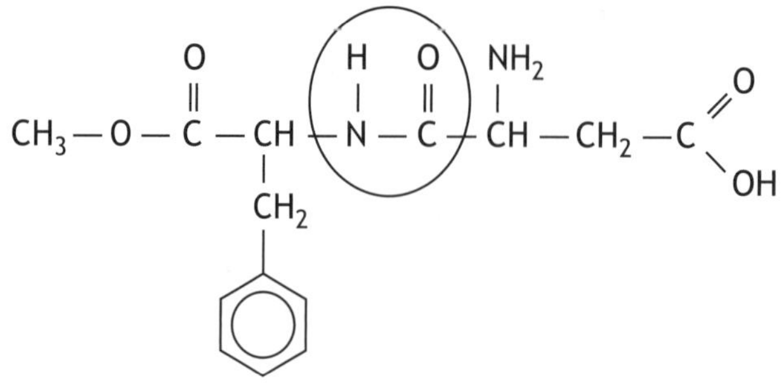

(i) Name the functional group circled. **1**

(ii) In the stomach, aspartame is hydrolysed by acid to produce methanol and two amino acids, phenylalanine and aspartic acid.

Two of the products of the hydrolysis of aspartame are shown below.

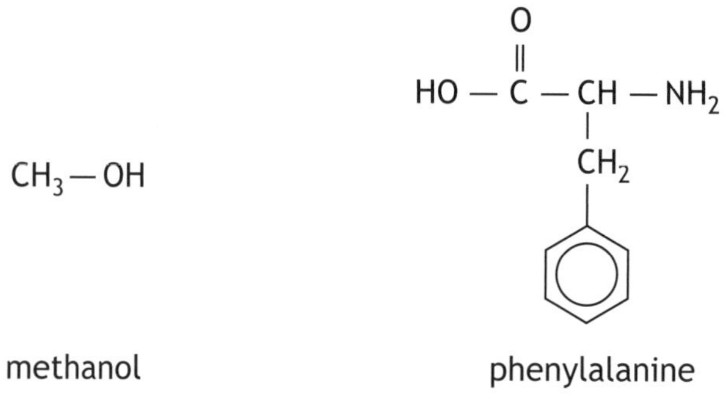

methanol

phenylalanine

Draw a structural formula for aspartic acid. **1**

(iii) The body cannot make all the amino acids it requires and is dependent on protein in the diet for the supply of certain amino acids.

What term is used to describe the amino acids the body cannot make? **1**

6. (continued)

(b) Caffeine is also added to some soft drinks. The concentration of caffeine can be found using chromatography.

A chromatogram for a standard solution containing 50 mg l^{-1} of caffeine is shown below.

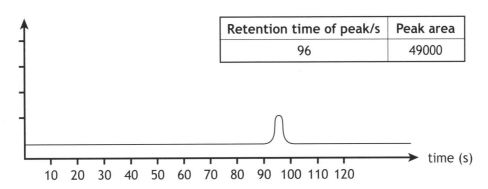

Retention time of peak/s	Peak area
96	49000

Results from four caffeine standard solutions were used to produce the calibration graph below.

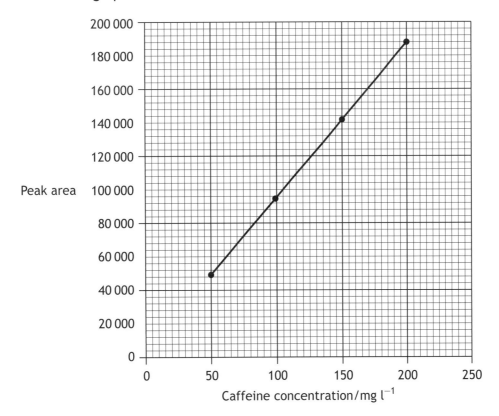

MARKS DO NOT WRITE IN THIS MARGIN

6. (b) (continued)

Chromatograms for two soft drinks are shown below.

Soft drink **X**

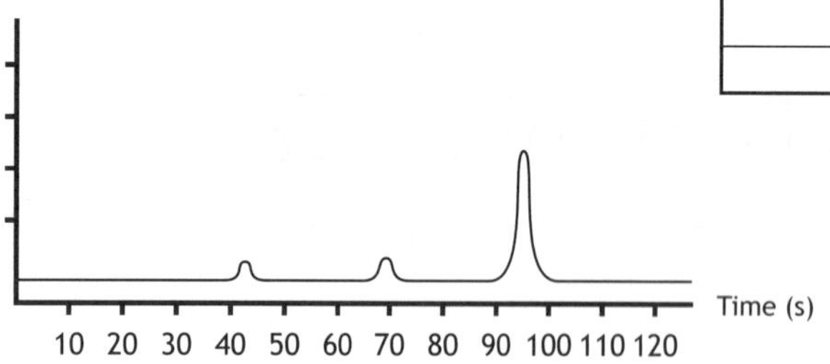

Retention time of peak/s	Peak area
42	11000
69	11350
96	68000

Soft drink **Y**

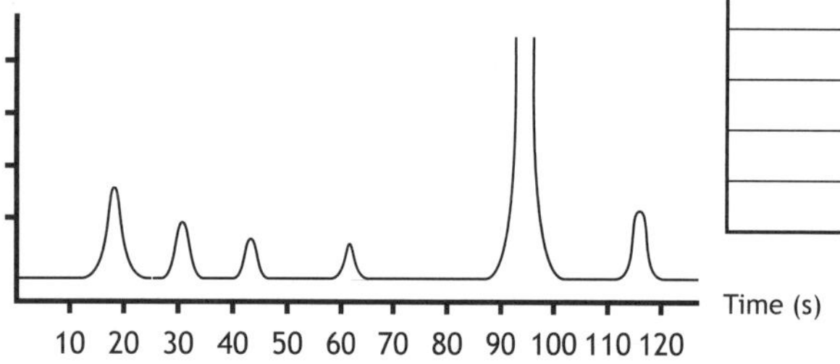

Retention time of peak/s	Peak area
17	7000
30	4600
43	3000
62	2500
96	——
115	5000

(i) What is the caffeine content, in mg l^{-1} of soft drink **X**? **1**

(ii) The caffeine content of the soft drink Y cannot be determined from its chromatogram.

What should be done to the sample of soft drink Y so that the caffeine content could be reliably calculated? **1**

Total marks 5

7. (a) A small sample of ammonia can be prepared in the laboratory by heating a mixture of ammonium chloride and calcium hydroxide. The ammonia is dried by passing it through small lumps of calcium oxide and collected by the downward displacement of air.

Complete the diagram to show how ammonia gas can be dried before collection.

1

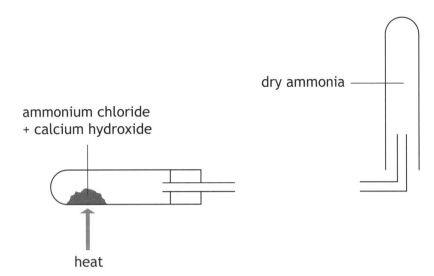

(b) Self-heating cans may be used to warm drinks such as coffee.

When the button on the can is pushed, a seal is broken, allowing water and calcium oxide to mix and react.

The reaction produces solid calcium hydroxide and releases heat.

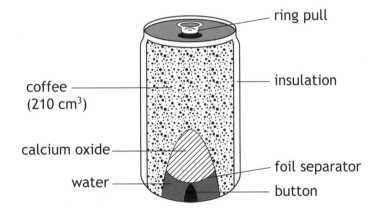

The equation for this reaction is:

$$CaO(s) + H_2O(\ell) \rightarrow Ca(OH)_2(s) \quad \Delta H = -65 \text{ kJ mol}^{-1}$$

7. (b) (continued)

(i) Calculate the mass, in grams, of calcium oxide required to raise the temperature of 210 cm³ of coffee from 20°C to 70°C. **3**

Show your working clearly.

(ii) If more water is used the calcium hydroxide is produced as a solution instead of as a solid.

The equation for the reaction is:

$$CaO(s) + H_2O(\ell) \rightarrow Ca(OH)_2(aq)$$

Using the following data, calculate the enthalpy change, in kJ mol^{-1}, for this reaction. **2**

$Ca(s) + \frac{1}{2}O_2(g) \rightarrow CaO(s)$ $\Delta H = -635$ kJ mol^{-1}

$H_2(g) + \frac{1}{2}O_2(g) \rightarrow H_2O(\ell)$ $\Delta H = -286$ kJ mol^{-1}

$Ca(s) + O_2(g) + H_2(g) \rightarrow Ca(OH)_2(s)$ $\Delta H = -986$ kJ mol^{-1}

$Ca(OH)_2(s) \rightarrow Ca(OH)_2(aq)$ $\Delta H = -82$ kJ mol^{-1}

Total marks   6

8. Benzoic acid, C_6H_5COOH, is an important feedstock in the manufacture of chemicals used in the food industry.

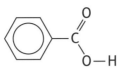

benzoic acid

(a) The ester ethyl benzoate is used as food flavouring.

Ethyl benzoate can be prepared in the laboratory by an esterification reaction. A mixture of ethanol and benzoic acid is heated, with a few drops of concentrated sulfuric acid added to catalyse the reaction.

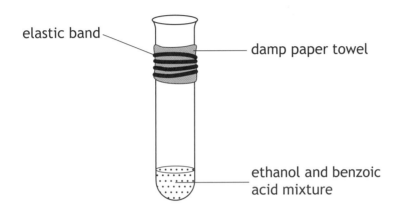

(i) Suggest a suitable method for heating the reaction mixture. 1

(ii) During esterification the reactant molecules join by eliminating a small molecule. What name is given to this type of chemical reaction? 1

(iii) Draw a structural formula for ethyl benzoate. 1

8. (continued)

(b) Sodium benzoate is used in the food industry as a preservative. It can be made by reacting benzoic acid with a concentrated solution of sodium carbonate.

$$2C_6H_5COOH + Na_2CO_3 \rightarrow 2C_6H_5COONa + CO_2 + H_2O$$

| mass of 1 mole = 122 g | mass of 1 mole = 106 g | mass of 1 mole = 144 g | mass of 1 mole = 44 g | mass of 1 mole = 18 g |

Calculate the atom economy for the production of sodium benzoate. **2**

(c) A chemist made 300g of sodium benzoate using 350g of benzoic acid and 280g of sodium carbonate.

The cost of the chemicals are shown below.

Benzoic acid	£15.80 for 100g
Sodium carbonate	£3.40 for 1kg

Calculate the cost of the chemicals required to produce 500g of sodium benzoate using this method. **2**

Total marks **7**

9. The table shows the boiling points of some alcohols.

Alcohol	Boiling point/°C
$CH_3CH_2CH_2CH_2OH$	118
$CH_3CH_2CH(OH)CH_3$	98
$CH_3CH(CH_3)CH_2OH$	108
$CH_3CH_2CH_2CH_2CH_2OH$	137
$CH_3CH_2CH_2CH(OH)CH_3$	119
$CH_3CH_2CH(CH_3)CH_2OH$	128
$CH_3CH_2C(OH)(CH_3)CH_3$	101
$CH_3CH_2CH_2CH_2CH_2CH_2OH$	159
$CH_3CH_2CH_2CH(CH_3)CH_2OH$	149
$CH_3CH_2CH_2C(OH)(CH_3)CH_3$	121

(a) Using information from the table, describe **two** ways in which differences in the structures affect boiling point of **isomeric alcohols**. 2

(b) Predict a boiling point for hexan-2-ol. 1

Total marks 3

10. Sherbet contains a mixture of sodium hydrogencarbonate and tartaric acid. The fizzing sensation in the mouth is due to the carbon dioxide produced in the following reaction.

$$2NaHCO_3 \;+\; C_4H_6O_6 \;\rightarrow\; Na_2(C_4H_4O_6) \;+\; 2H_2O \;+\; 2CO_2$$

sodium hydrogencarbonate — tartaric acid — sodium tartrate

(a) The chemical name for tartaric acid is 2,3-dihydroxybutanedioic acid.

Draw a structural formula for tartaric acid. **1**

(b) In an experiment, a student found that adding water to 20 sherbet sweets produced 105 cm³ of carbon dioxide.

Assuming that sodium hydrogencarbonate is in excess, calculate the average mass of tartaric acid, in grams, in one sweet. **3**

(Take the molar volume of carbon dioxide to be 24 litre mol⁻¹.)

Show your working clearly.

Total marks **4**

11. When cyclopropane gas is heated over a catalyst, it isomerises to form propene gas and an equilibrium is obtained.

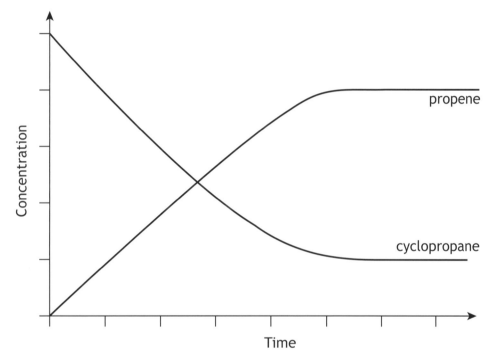

The graph shows the concentration of cyclopropane and propene as equilibrium is established in the reaction.

(a) Mark clearly on the graph the point at which equilibrium has just been reached. **1**

(b) Why does increasing the pressure have no effect on the position of this equilibrium. **1**

11. (continued)

(c) The equilibrium can also be achieved by starting with propene.

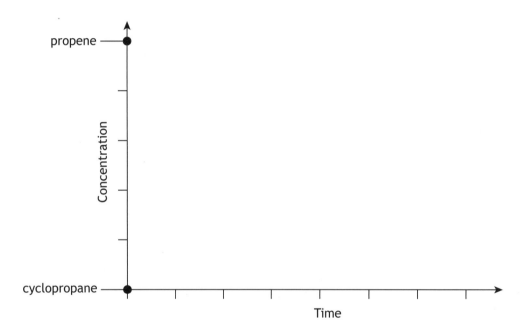

Using the initial concentrations shown, sketch a graph to show how the concentrations of propene and cyclopropane change as equilibrium is reached for this reverse reacton.

1

Total marks 3

12. Aldehydes and ketones can take part in a reaction sometimes known as an aldol condensation.

    The simplest aldol reacion involves two molecules of ethanal.

    ```
         H   H              H   H                    H   H   H   H
         |   |              |   |                    |   |   |   |
    H  - C - C = O  +  H -  C - C = O    ———>   H -  C - C - C - C = O
         |                  |                        |   |   |
         H                  H                        H   OH  H
    ```

    In the reaction, the carbon atom next to the carbonyl functional group of one molecule forms a bond with the carbonyl carbon atom of the second molecule.

    (a) Draw a structural formula for the product formed when propanone is used instead of ethanal in this type of reaction. **1**

    (b) Name an aldehyde that would **not** take part in an aldol condensation. **1**

    (c) Apart from the structure of the reactants, suggest what is unusual about applying the term "condensation" to this particular type of reaction. **1**

    Total marks   3

13. Oxalic acid is found in rhubarb. The number of moles of oxalic acid in a carton of rhubarb juice can be found by titrating samples of the juice with a solution of potassium permanganate, a powerful oxidising agent.

    The equation for the overall reaction is:

    $$6(COOH)_2(aq) + 6H^+(aq) + 2MnO_4^-(aq) \rightarrow 2Mn^{2+}(aq) + 10CO_2(aq) + 8H_2O(\ell)$$

    (a) Write the ion-electron equation for the reduction reaction. **1**

    (b) In an investigation using a 500 cm³ carton of rhubarb juice, separate 25·0 cm³ samples were measured out. Three samples were then titrated with 0·040 mol l⁻¹ potassium permanganate solution, giving the following results.

Titration	Volume of potassium permanganate solution used/cm³
1	27·7
2	26·8
3	27·0

    Average volume of potassium permanganate solution used = 26·9cm³.

    (i) Why was the first titration result not included in calculating the average volume of potassium permanganate solution used? **1**

    (ii) Suggest a suitable piece of apparatus that could be used to measure out the 25cm³. **1**

    (iii) Calculate the mass of oxalic acid present in the 500cm³ carton of rhubarb juice. **4**

    Show your working clearly.

Total marks **7**

14. A fatty acid is a long chain carboxylic acid.

    Examples of fatty acids are shown in the table below.

Common name	Systemic name	Structure
stearic acid	octadecanoic acid	$CH_3(CH_2)_{16}COOH$
oleic acid	octadec-9-enoic acid	$CH_3(CH_2)_7CH=CH(CH_2)_7COOH$
linoleic acid	octadec-9,12-dienoic acid	$CH_3(CH_2)_4CH=CHCH_2CH=CH(CH_2)_7COOH$
linolenic acid		$CH_3CH_2CH=CHCH_2CH=CHCH_2CH=CH(CH_2)_7COOH$

    (a) What is the systematic name for linolenic acid? **1**

    (b) Stearic acid reacts with sodium hydroxide solution to form sodium stearate.

    sodium stearate

    (i) Name the type of reaction taking place when stearic acid reacts with sodium hydroxide. **1**

    (ii) **Explain fully** how sodium stearate acts to keep grease and non-polar substances suspended in water during cleaning. **3**

    Total marks **5**

15. Infra-red spectroscopy is a technique that can be used to identify the bonds that are present in a molecule.

Different bonds absorb infra-red radiation of different wavenumbers. This is due to differences in the bond 'stretch'. These absorptions are recorded in a spectrum.

A spectrum for propan-1-ol is shown.

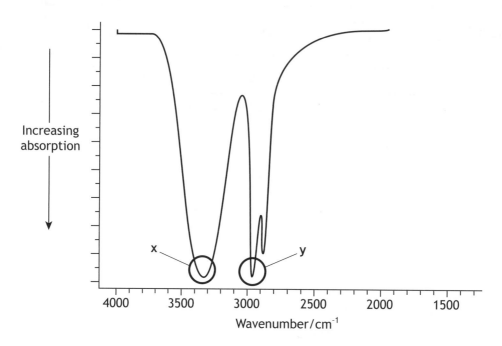

The correlation table on page 14 of the data booklet shows the wavenumber ranges for the absorptions due to different bonds.

(a) Use the correlation table to identify the bonds responsible for the two absorptions, **x** and **y**, that are circled in the propan-1-ol spectrum.

x: y:

(b) Propan-1-ol reacts with ethanoic acid.

Draw a spectrum that could be obtained for the organic product of this reaction.

1

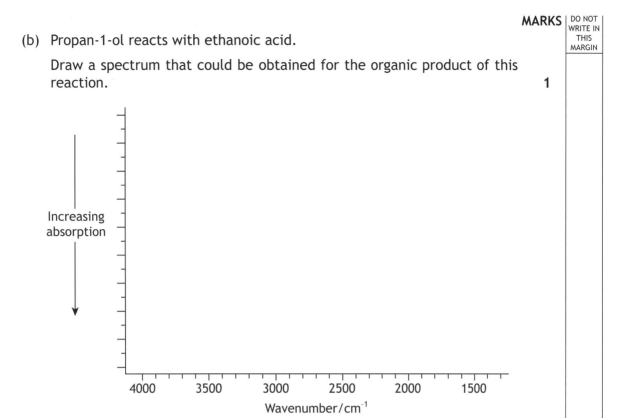

Total marks 2

[END OF MODEL PAPER]

ADDITIONAL SPACE FOR ANSWERS AND ROUGH WORK

ADDITIONAL SPACE FOR ANSWERS AND ROUGH WORK

ADDITIONAL SPACE FOR ANSWERS AND ROUGH WORK

HIGHER FOR CfE

Model Paper 2

Whilst this Model Paper has been specially commissioned by Hodder Gibson for use as practice for the Higher (for Curriculum for Excellence) exams, the key reference documents remain the SQA Specimen Paper 2014 and SQA Past Paper 2015.

National Qualifications
MODEL PAPER 2

**Chemistry
Section 1 — Questions**

Duration — 2 hours and 30 minutes

Reference may be made to the Chemistry Higher and Advanced Higher Data Booklet.

Instructions for the completion of Section 1 are given on *Page two* of your question and answer booklet.

Record your answers on the answer grid on *Page three* of your question and answer booklet.

Before leaving the examination room you must give your question and answer booklet to the Invigilator; if you do not you may lose all the marks for this paper.

SECTION 1 — 20 marks
Attempt ALL questions

1. Which of the following changes could lead to an increase in reaction rate?

   A increase in particle size
   B increase in activation energy
   C increase in pressure
   D increase in enthalpy change

2. The energy profile shown is for an exothermic reaction.

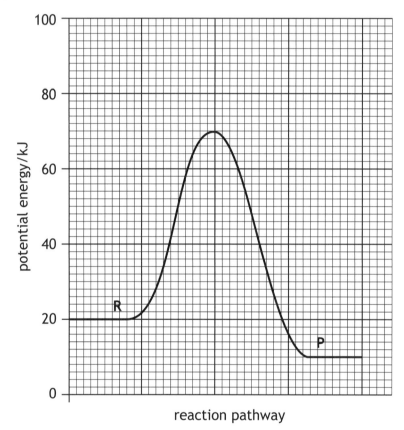

Key: R = reactants   P = products

What is the activation energy, in kJmol^{-1}, for the forward reaction?

   A 50
   B -10
   C 10
   D 60

3. The melting point of the halogens, group 7, is shown in the table.

Halogen	Melting point, °C
Fluorine	−220
Chlorine	−101
Bromine	−7
Iodine	114

Which of the following is **not** a correct explanation for this trend in melting point?

A The molecules with most electrons have the strongest intermolecular forces.

B The London dispersion force increases as the molecular mass increases.

C The melting point increases as more energy is required to break the strong covalent bonds.

D The force of attraction between molecules increases going down the group.

4. The energy change for the reaction $Al(g) \rightarrow Al^{3+}(g)$ is equal to

A the 3rd ionisation energy of Al

B the 1st + 3rd ionisation energy of Al

C the 3rd - (2nd + 1st) ionisation energy of Al

D the 1st + 2nd + 3rd ionisation energy of Al.

5. Which of the following statements correctly explains the finding that water has a higher boiling point than ethane, C_2H_6.

A The covalent bonds in water are stronger than the covalent bonds in ethane.

B The hydrogen bonds between water molecules are stronger than the hydrogen bonds between ethane molecules.

C The hydrogen bonds between water molecules are stronger than the London dispersion forces between ethane molecules.

D The polar O-H bond is harder to break than the C-H bond.

6. Which of the following bonding types could be found in an element?

    A   Hydrogen bonds
    B   Polar covalent bonds
    C   London dispersion forces
    D   Ionic bonds

7. The strength of van der Waals' forces would be

    A   London dispersion forces < dipole-dipole < hydrogen bonding
    B   hydrogen bonding < London dispersion forces < dipole-dipole
    C   dipole-dipole < London dispersion forces < hydrogen bonding
    D   London dispersion forces < hydrogen bonding < dipole-dipole.

8. Which line in the table correctly describes the reaction between an alcohol and carboxylic acid to form an ester?

	Description of reaction
A	The OH from the alcohol reacts with any H from the acid to form an ester and water
B	The OH from the carboxylic acid reacts with the H from the hydroxyl to form water. An ester is also formed.
C	A condensation reaction takes place between the hydroxide group of the alcohol and carboxylic group to form an ester and water.
D	The OH from the alcohol reacts with an H from the carboxylic acid to form an ester and water in a hydrolysis reaction.

9. Which of the following statements would not apply to a protein?

    A   They can be hydrolysed to form amino acids.
    B   They can react with alkalis to form soaps.
    C   They contain the amide group.
    D   They can hydrogen bond to other protein chains.

10. The compound shown below is an example of

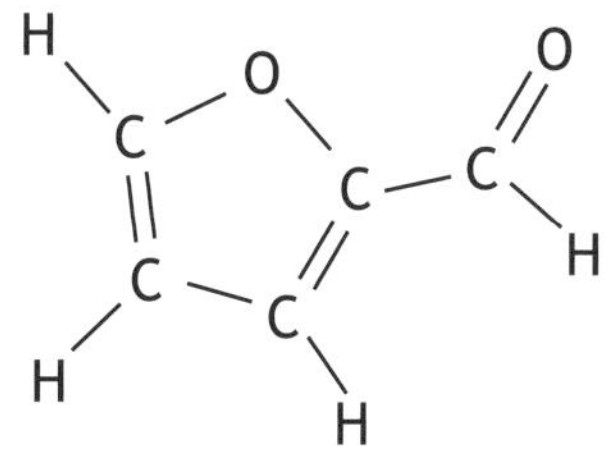

    A   a primary alcohol
    B   a secondary alcohol
    C   an aldehyde
    D   a ketone.

11. Which of the following compounds could react with acidified potassium dichromate to form an acidic compound?

    A   propan-2-ol
    B   2-methyl propan-2-ol
    C   propanone
    D   propanal

12. Vanillin is the flavour molecule found in vanilla.

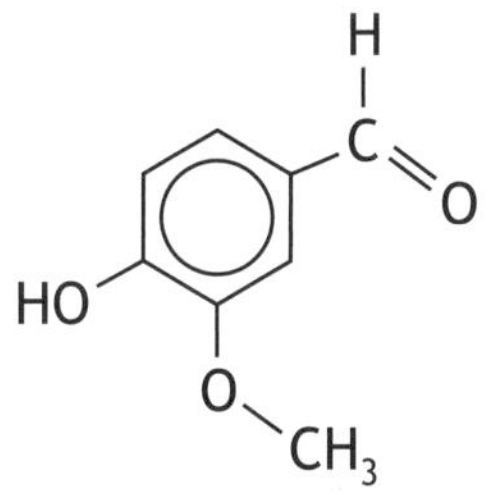

    Which of the following statements could be applied to vanillin?

    A   It can be oxidised to form a carboxylic acid.
    B   It will be insoluble in water.
    C   It is an ester.
    D   It is a ketone.

13. Cl· + Cl· → Cl₂

    The above reaction can be described as a

    A   termination
    B   propagation
    C   initiation
    D   oxidation.

14. Mg(s) + O₂(g) → MgO(s)

    The mass of MgO produced from reacting 2.43g of Mg with excess oxygen is

    A   4.03g
    B   40.3g
    C   18.43g
    D   34.43g.

15. Br₂(l) + H₂O(l) ⇌ Br⁻(aq) + BrO⁻(aq) + 2H⁺(aq)

    Which of the following, when added to the reaction at equilibrium, will **not** cause a shift to the right?

    A   Br₂(l)
    B   NaOH(aq)
    C   HCl(aq)
    D   CaCO₃(s)

16. The enthalpy of combustion of ethanol is $-1367\ kJmol^{-1}$.

    The energy released when 2.3g of ethanol burns is

    A   68.35kJ
    B   27340kJ
    C   68.35kJmol⁻¹
    D   27340kJmol⁻¹

17. Mg(s) + 2H⁺(aq) → Mg²⁺(aq) + H₂(g)   ΔH = a

Zn(s) + 2H⁺(aq) → Zn²⁺(aq) + H₂(g)   ΔH = b

Mg(s) + Zn²⁺(aq) → Mg²⁺(aq) + Zn(s)   ΔH = c

According to Hess' Law

- A   c = a−b
- B   c = a+b
- C   c = b−a
- D   c = −b−a.

18. In which of the following reactions is hydrogen acting as an oxidising agent?

- A   Li + H₂ → 2LiH
- B   ZnO + H₂ → Zn + H₂O
- C   F₂ + H₂ → 2HF
- D   C₃H₆ + H₂ → C₃H₈

19. Oxidising agents can be used to

- A   extract metal elements from metal compounds
- B   destroy bacteria
- C   terminate free radical reactions
- D   absorb UV light.

20. A mixture of propan-1-ol, propanal and propanone was passed through a chromatography column. The table below shows the retention times for the compounds.

Compound	Retention Time, s
Propan-1-ol	45
Propanone	12
Propanal	13

   The retention time for propane-1,2,3-triol would be

   A more than 45s

   B less than 12s

   C between 12 and 13s

   D between 12 and 45s.

[END OF SECTION 1. NOW ATTEMPT THE QUESTIONS IN SECTION 2 OF YOUR QUESTION AND ANSWER BOOKLET.]

National Qualifications
MODEL PAPER 2

Chemistry
Section 1 — Answer Grid and Section 2

Duration — 2 hours and 30 minutes

Reference may be made to the Chemistry Higher and Advanced Higher Data Booklet.

Total marks — 100

SECTION 1 — 20 marks

Attempt ALL questions.

Instructions for completion of Section 1 are given on *Page 104*.

SECTION 2 — 80 marks

Attempt ALL questions

Write your answers clearly in the spaces provided in this booklet. Additional space for answers and rough work is provided at the end of this booklet. If you use this space you must clearly identify the question number you are attempting. Any rough work must be written in this booklet. You should score through your rough work when you have written your final copy.

Use **blue** or **black** ink.

Before leaving the examination room you must give this booklet to the Invigilator; if you do not you may lose all the marks for this paper.

SECTION 1— 20 marks

The questions for Section 1 are contained on *Page 95* — Questions.
Read these and record your answers on the answer grid on *Page 105* opposite.
DO NOT use gel pens.

1. The answer to each question is **either** A, B, C or D. Decide what your answer is, then fill in the appropriate bubble (see sample question below).

2. There is **only one correct** answer to each question.

3. Any rough working should be done on the additional space for answers and rough work at the end of this booklet.

Sample Question

To show that the ink in a ball-pen consists of a mixture of dyes, the method of separation would be:

    A   fractional distillation

    B   chromatography

    C   fractional crystallisation

    D   filtration.

The correct answer is **B**—chromatography. The answer **B** bubble has been clearly filled in (see below).

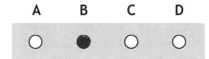

Changing an answer

If you decide to change your answer, cancel your first answer by putting a cross through it (see below) and fill in the answer you want. The answer below has been changed to **D**.

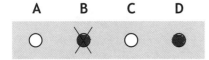

If you then decide to change back to an answer you have already scored out, put a tick (✓) to the **right** of the answer you want, as shown below:

SECTION 1 — Answer Grid

	A	B	C	D
1	○	○	○	○
2	○	○	○	○
3	○	○	○	○
4	○	○	○	○
5	○	○	○	○
6	○	○	○	○
7	○	○	○	○
8	○	○	○	○
9	○	○	○	○
10	○	○	○	○
11	○	○	○	○
12	○	○	○	○
13	○	○	○	○
14	○	○	○	○
15	○	○	○	○
16	○	○	○	○
17	○	○	○	○
18	○	○	○	○
19	○	○	○	○
20	○	○	○	○

SECTION 2 — 80 marks

Attempt ALL questions

1. The Periodic Table allows chemists to make predictions about the properties of elements.

   (a) The elements lithium to neon make up the second period of the Periodic Table.

Li	Be	B	C	N	O	F	Ne

   (i) Name an element from the second period that exists as a covalent network. **1**

   (ii) Describe the structure of Neon. **1**

   (iii) Which element in the second period is the stongest reducing agent? **1**

   (b) Compounds of the group 1 elements are widely used.

   Potassium nitrate produces oxygen gas when heated.

   $KNO_3(s) \rightarrow KNO_2(s) + \frac{1}{2}O_2(g)$

   Calculate the mass of potassium nitrate required to produce 3.8 litres of oxygen gas. **3**

   (Take the molar volume of oxygen to be 24 litres mol^{-1})

Total marks **6**

2. Trends in the periodic table such as ionisation energy, electronegativity and covalent radius can be explained with a basic knowledge of the structure of the atom. **Using your knowledge of chemistry**, discuss this statement.

3

3. A student carried out three experiments involving the reaction of excess magnesium ribbon with dilute acids. The rate of hydrogen production was measured in each of the three experiments.

Experiment	Acid
1	100 cm^3 of 0·10 mol l^{-1} sulfuric acid
2	50 cm^3 of 0·20 mol l^{-1} sulfuric acid
3	100 cm^3 of 0·10 mol l^{-1} hydrochloric acid

The equation for **Experiment 1** is shown.

$$Mg(s) + H_2SO_4(aq) \rightarrow MgSO_4(aq) + H_2(g)$$

(a) The curve obtained for **Experiment 1** is drawn on the graph.

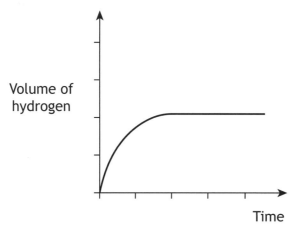

Draw curves on the graph to show the results obtained for **Experiment 2** and **Experiment 3**.

Label each curve clearly.

(b) The graph shows the distribution of kinetic energy for molecules in a reaction mixture at a given temperature.

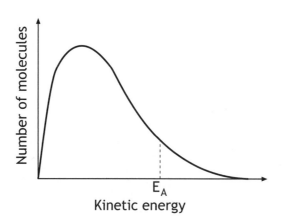

Why does a small increase in temperature produce a large increase in reaction rate.

Total marks 6

MARKS | DO NOT WRITE IN THIS MARGIN

4. (a) Hydrogen and fluorine can react explosively to form hydrogen fluoride gas.

The equation for the reaction is shown.

$$H_2(g) \ + \ F_2(g) \rightarrow 2HF(g)$$

Using bond enthalpy values from the data booklet, caluculate the enthalpy change for this reaction. **2**

(b) The boiling point of hydrogen fluoride, HF, is much higher than the boiling point of F_2.

H —— F	F —— F
boiling point: 19·5 °C	boiling point: −188 °C

Explain fully why the boiling the point of hydrogen fluoride is much higher than the boiling point of fluorine.

In your answer you should mention the intermolecular forces involved and how they arise. **3**

Total marks 5

5. A standard solution of iodine can be used to determine the mass of vitamin C in orange juice.

   Iodine reacts with vitamin C, $C_6H_8O_6$, as shown by the following redox equation.

   $$C_6H_8O_6(aq) + I_2(aq) \rightarrow C_6H_6O_6(aq) + 2H^+(aq) + 2I^-(aq)$$

   In an investigation using a carton containing 500 cm³ of orange juice, 50 cm³ samples were measured out using a measuring cylinder.

   Each sample was then titrated with a 0.0050 mol l⁻¹ solution of iodine. The results of the titrations are shown in the table.

Titration	Volume of iodine, cm³
1	23.0
2	22.3
3	22.5

   (a) Titrating the whole carton of orange juice would require large volumes of iodine solution.

   Apart from this disadvantage give another reason for titrating several smaller samples of orange juice. **1**

   (b) An average of 22·4 cm³ of 0.0050 mol l⁻¹ iodine solution was required for the complete titration of the vitamin C in a 50·0 cm³ sample of orange juice.

   Calculate the mass, in grams, of vitamin C in the 500 cm³ carton of orange juice. **3**

   (mass of 1 mole vitamin C = 176 g)

   Show your working clearly.

MARKS | DO NOT WRITE IN THIS MARGIN

5. (continued)

(c) The student carrying out this experiment found that their results were not accurate.

What mistake did the student make when carrying out this experiment? **1**

(d) Some vitamin C tablets contain a compound of vitamin C known as Ester-C. The structures for vitamin C and ester C are shown below.

Vitamin C structure

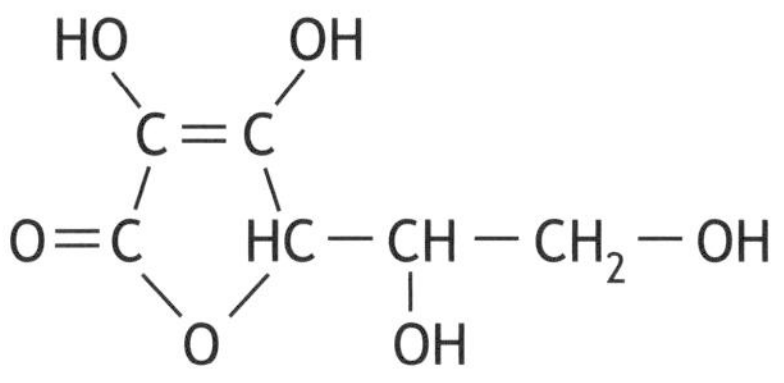

Ester-C structure

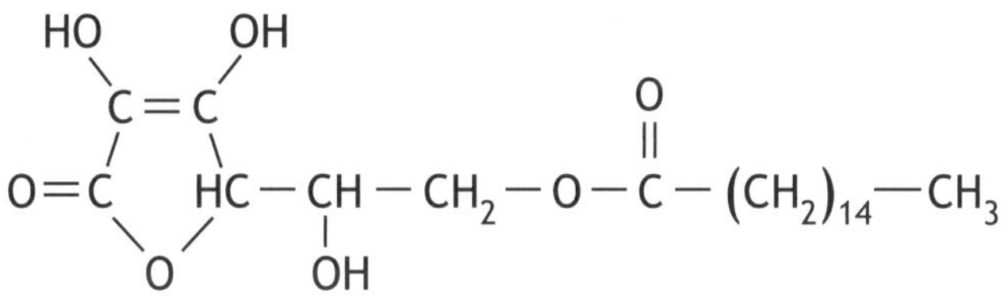

Ester-C breaks down in the body to form vitamin C and a carboxylic acid.

(i) Name the chemical reaction which results in the Ester-C forming vitamin C and a carboxylic acid. **1**

(ii) Why is vitamin C more soluble in water than Ester-C? **1**

(e) The recommended daily allowance of vitamin C is 60 mg.

50 cm^3 of a fruit juice contained 7.3 mg of vitamin C.

Calculate the volume of fruit juice which would provide the recommended daily allowance of vitamin C. **2**

Total marks 9

MARKS | DO NOT WRITE IN THIS MARGIN

6. Chocolate contains various compounds.

(a) Many of the flavour and aroma molecules found in chocolate are aldehydes and ketones.

Two examples are shown below.

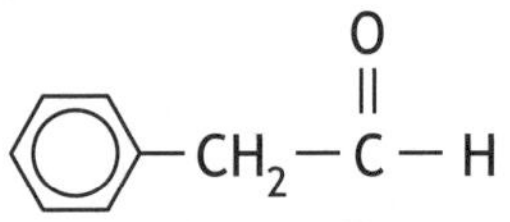

phenylethanal

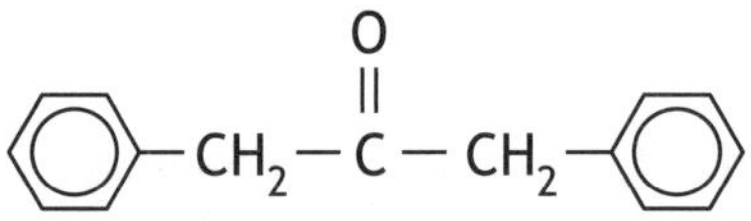

1,3-diphenylpropan-2-one

Phenylethanal can be easily oxidised but 1,3-diphenylpropan-2-one cannot.

(i) Name a chemical that could be used to distinguish between these two compounds. **1**

(ii) Name the type of organic compound formed when phenylethanal is oxidised. **1**

(b) Glycerol monostearate is an emulsifier used in chocolate.

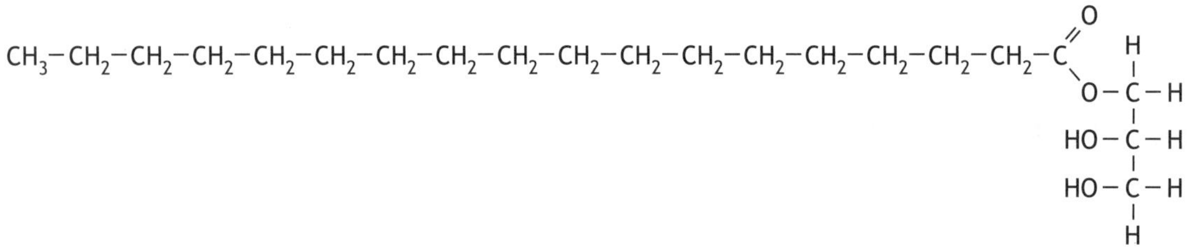

(i) Why is glycerol monostearate added to chocolate? **1**

(ii) Draw a structural formula for glycerol. **1**

MARKS DO NOT WRITE IN THIS MARGIN

6. (continued)

(c) Theobromine, a compound present in chocolate, can cause illness in dogs and cats.

To decide if treatment is necessary, vets must calculate the mass of theobromine consumed.

1·0 g of chocolate contains 1·4 mg of theobromine.

Calculate the mass, in mg, of theobromine in a 17 g biscuit of which 28% is chocolate. **2**

Show your working clearly.

(d) The flavour and texture of chocolate comes from a blend of compounds.

Using your knowledge of chemistry, describe how you could show that there are ionic compounds and covalent compounds present in chocolate. **3**

Total marks 9

7. Dental anaesthetics are substances used to reduce discomfort during treatment.

(a) Lidocaine is a dental anaesthetic.

Lidocaine causes numbness when applied to the gums. This effect wears off as the lidocaine is hydrolysed.

One of the products of the hydrolysis of lidocaine is compound **C**.

compound **C**

(i) Name the functional group circled above. **1**

(ii) Draw a structural formula for the other compound produced when lidocaine is hydrolysed. **1**

(iii) Draw a structural formula for the organic compound formed when compound **C** reacts with NaOH(aq). **1**

7. (continued)

(b) The table below shows the duration on numbness for common anaesthetics.

Name of anaesthetic	Structure	Duration of numbness/ minutes
procaine	(structure)	7
lidocaine	(structure)	96
mepivacaine	(structure)	114
anaesthetic X	(structure)	

Estimate the duration of numbness, in minutes, for anaesthetic X.

7. (continued)

(c) The maximum safe dose of lidocaine for an adult is 4.5 mg of lidocaine per kg of body mass.

1.0 cm³ of lidocaine solution contains 10mg of lidocaine.

Calculate the maximum volume of lidocaine solution that could be given to a 70kg adult.

Show your working clearly.

3

(d) When forensic scientists analyse illegal drugs, anaesthetics such as lidocaine are sometimes found to be present.

The gas chromatogram below is from an illegal drug.

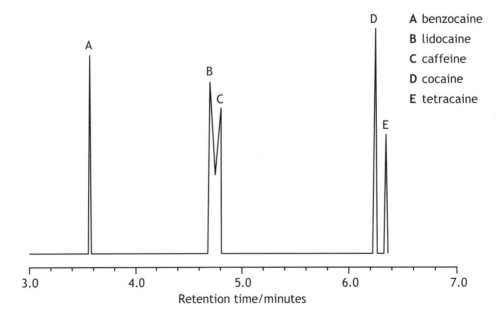

A benzocaine
B lidocaine
C caffeine
D cocaine
E tetracaine

MARKS | DO NOT WRITE IN THIS MARGIN

7. (d) (continued)

(i) The structures of benzocaine and tetracaine are shown below.

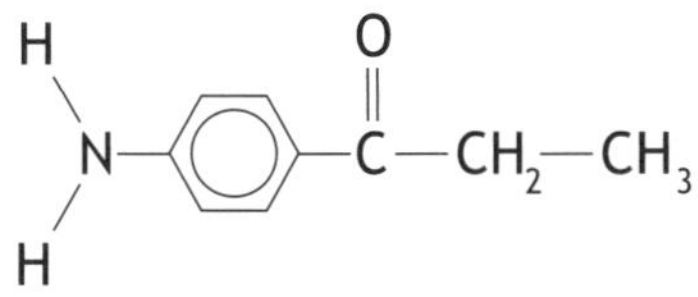

benzocaine

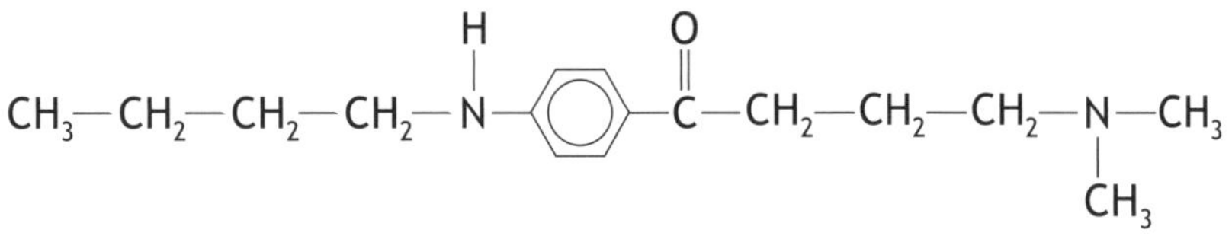

tetracaine

Suggest why benzocaine has a shorter retention time than tetracaine. 1

(ii) Why is it difficult to obtain accurate values for the amount of lidocaine present in a sample containing large amounts of caffeine?

1

(iii) Add a peak to the diagram below to complete the chromatogram for a second sample that only contains half the amount of tetracaine compared to the first.

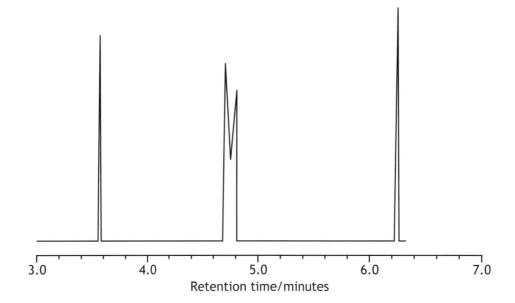

1

Total marks 10

8. Ozone, O_3, can be destroyed when it reacts with chlorine free radicals.

   The chlorine free radicals are formed when CFCs, such as CCl_2F_2, react in the upper atmosphere. The first step of the reaction is shown by the equation below.

   $$CCl_2F_2 \rightarrow Cl^\bullet + {}^\bullet CClF_2$$

   (a) Draw a structural formula for CCl_2F_2

   (b) Why do free radicals form in the upper atmosphere?

   (c) Chlorine free radicals react with ozone molecules to form oxygen gas.

   $$Cl^\bullet + O_3 \rightarrow ClO^\bullet + O_2$$

   (i) What name is given to this type of reaction?

   (ii) Explain how a free radical scavenger could prevent ozone being destroyed by CFCs in the upper atmosphere.

   Total marks 5

9. Carbon monoxide can be produced in many ways.

   (a) One method involves the reaction of carbon with an oxide of boron.

   $$B_2O_3 + C \rightarrow B_4C + CO$$

   Balance this equation.

   (b) Carbon monoxide is also a product of the reaction of carbon dioxide with hot carbon. The carbon dioxide is made by the reaction of dilute hydrochloric acid with solid calcium carbonate.

   Unreacted carbon dioxide is removed before the carbon monoxide is collected by displacement of water.

   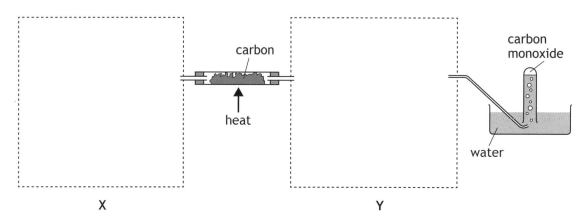

   Complete the diagram to show how the carbon dioxide can be produced at **X** and how the unreacted carbon dioxide can be removed by bubbling it through a solution at **Y**.

   Normal laboratory apparatus should be used in your answer and the chemicals used at **X** and **Y** should be labelled.

   (c) Carbon monoxide is a useful reducing agent. It is used to form iron metal from iron (III) oxide.

   $$Fe_2O_3(s) + 3CO(g) \rightarrow 2Fe(s) + 3CO_2(g)$$

   Calculate the atom economy for the production of iron.

   Total marks 5

MARKS DO NOT WRITE IN THIS MARGIN

10. Paracetamol is a widely used painkiller.

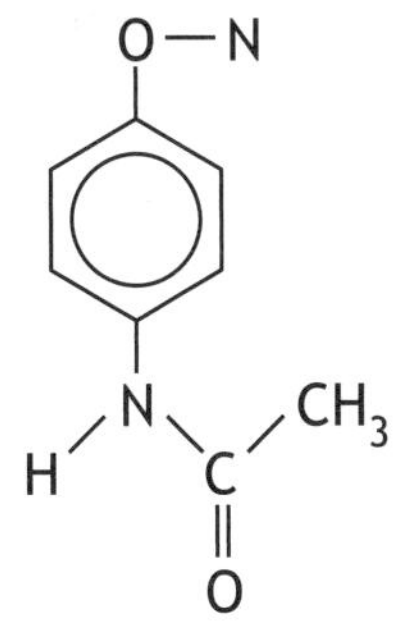

(a) Circle the amide link in the structure shown above. **1**

(b) Paracetamol can be made by the reaction 4-aminophenol with ethanoic anhydride.

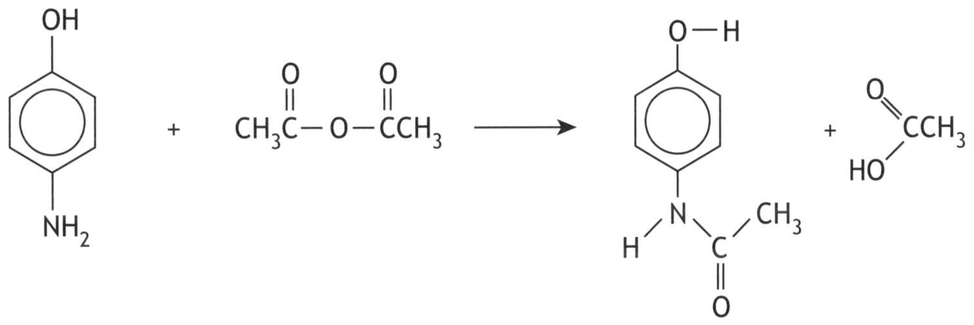

4-aminophenol (mass of 1mole = 109g)	ethanoic anhydride	paracetamol (mass of 1mole = 151g)	ethanoic acid

3.2g of paracetamol was produced when 4.5g of 4-aminophenol was reacted with an excess of ethanoic anhydride.

(i) Calculate the percentage yield. **2**

Show your working clearly.

MARKS | DO NOT WRITE IN THIS MARGIN

10. (b) (continued)

(ii) Calculate the cost of the 4-aminophenol required to produce 500g of paracetamol using this method. **2**

The cost of 4-aminophenol is shown below.

4-aminophenol	£57.20 for 1kg

(c) One antidote for paracetamol overdose is methionine.

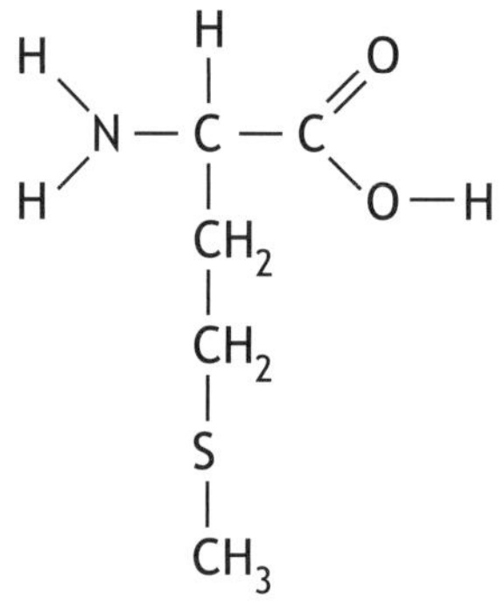

To what family of organic compounds does methionine belong? **1**

10. (continued)

(d) The concentration of paracetamol in a solution can be determined by measuring how much UV radiation it absorbs.

The graph shows how the absorbance of a sample containing 0·040 g l^{-1} paracetamol varies with wavelength.

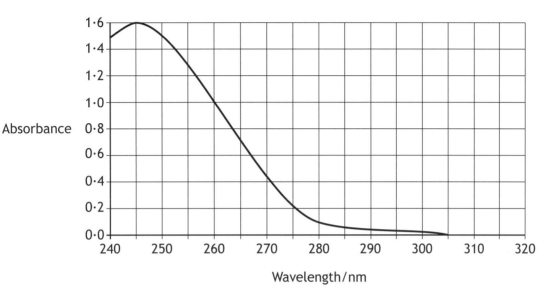

The quantity of UV radiation of wavelength 245 nm absorbed is directly proportional to the concentration of paracetamol.

The absorbance of a second sample of paracetamol solution measured at 245 nm was 0·90.

What is the concentration, in g l^{-1}, of this second paracetamol solution? **1**

(e) Describe how a solution of 0.040 gl^{-1} paracetamol solution would be prepared. **2**

Total marks 9

11. The structure of a molecule found in olive oil can be represented as shown.

(a) Olive oil can be hydrolysed using sodium hydroxide solution to produce sodium salts of fatty acids.

Name the other product of this reaction. **1**

(b) In what way does the structure of a fat molecule differ from that of an oil molecule? **1**

(c) Over time, open containers of olive oil develop a rancid flavour. What substance is reacting with the oil to cause these unwanted changes to take place? **1**

Total marks **3**

12. Chemists have developed cheeses specifically for use in cheeseburgers.

(a) When ordinary cheddar cheese is grilled the shapes of the protein molecules change and the proteins and fats separate leaving a chewy solid and an oily liquid.

What name is given to the change in protein structure which occurs when ordinary cheddar is grilled? **1**

(b) To make cheese for burgers, grated cheddar cheese, soluble milk proteins and some water are mixed and heated to no more than 82°C. As the cheese begins to melt an emulsifying agent is added and the mixture is stirred.

(i) Why would a water bath be used to heat the mixture? **1**

(ii) A section of the structure of a soluble milk protein is shown below.

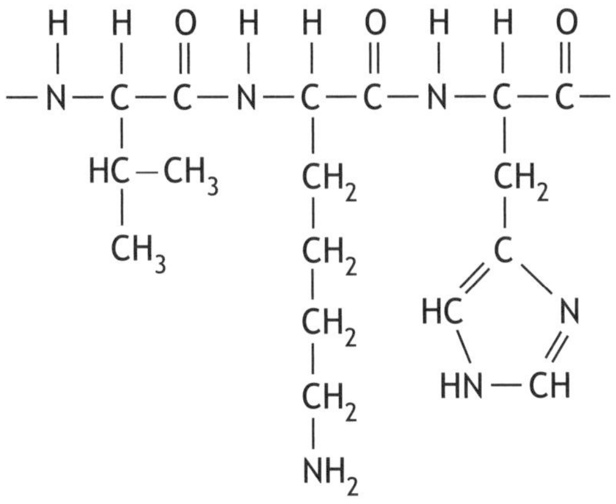

$$-N-C-C-N-C-C-N-C-C-$$

with side chains: $HC-CH_3$ / CH_3 ; $(CH_2)_4-NH_2$; CH_2-imidazole ring

Draw a structural formula for any **one** of the amino acids formed when this section of protein is hydrolysed. **1**

12. (b) (continued)

    (iii) The emulsifier used is trisodium citrate, a salt formed when citric acid is neutralised using sodium hydroxide.

    Complete the equation below showing a structural formula for the trisodium citrate formed.

1

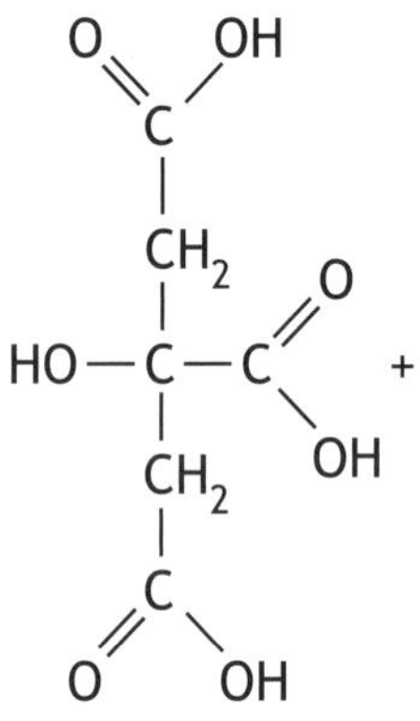

$$\text{structure of citric acid} + 3\text{NaOH} \longrightarrow + 3H_2O$$

(Citric acid structure shown: central C bonded to OH, CH₂COOH (top), COOH, and CH₂COOH (bottom))

Total marks   4

13. Mobile phones are being developed that can be powered by methanol.

Methanol can be made by a two-stage process.

(a) In the first stage, methane is reacted with steam to produce a mixture of carbon monoxide and hydrogen.

$$CH_4(g) + H_2O(g) \rightleftharpoons CO(g) + 3H_2(g)$$

Use the data below to calculate the enthalpy change, in kJ mol^{-1}, for the forward reaction.

$CO(g) + \frac{1}{2}O_2(g) \rightarrow CO_2(g)$ $\quad\quad \Delta H = -283$ kJ mol^{-1}

$H_2(g) + \frac{1}{2}O_2(g) \rightarrow H_2O(g)$ $\quad\quad \Delta H = -242$ kJ mol^{-1}

$CH_4(g) + 2O_2(g) \rightarrow CO_2(g) + 2H_2O(g)$ $\quad\quad \Delta H = -803$ kJ mol^{-1}

Show your working clearly.

(b) In the second stage, the carbon monoxide and hydrogen react to produce methanol.

$$CO(g) + 2H_2(g) \rightleftharpoons CH_3OH(g) \quad \Delta H = -91 \text{ kJ mol}^{-1}$$

Circle the correct words in the table to show the changes to temperature and pressure that would favour the production of methanol.

temperature	decrease / keep the same / increase
pressure	decrease / keep the same / increase

13. (continued)

(c) A student set-up the following apparatus and carried out an experiment to determine the enthalpy of combustion of methanol.

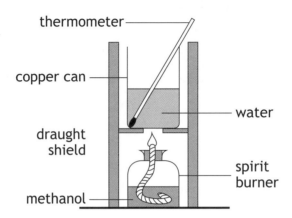

The results of the student's experiment are shown in the table below.

Mass of methanol burned	0.38g
Volume of water heated	100cm^3
Temperature of water at the start	21.7°C
Higher temperature of water after heating	37.8°C

Use the student's experimental results to calculate the enthalpy of combustion of methanol.

3

MARKS DO NOT WRITE IN THIS MARGIN

13. (continued)

(d) A more accurate value can be obtained using a bomb calorimeter.

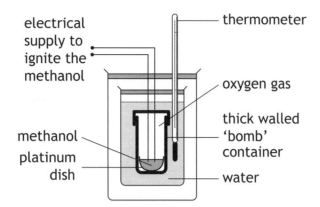

One reason for the more accurate value is that less heat is lost to the surroundings than in the simple laboratory method.

Give one other reason for the value being more accurate in the bomb calorimeter method. 1

Total marks 7

14. Carbon-13 NMR is a technique used in chemistry to determine the structure of organic compounds.

    The technique allows a carbon atom in a molecule to be identified by its 'chemical shift'. This value depends on the other atoms bonded to the carbon atom.

    Shift table

Carbon environment	Chemical shift/ppm
C = O (in ketones)	205 – 220
C = O (in aldehydes)	190 – 205
C = O (in acids and esters)	170 – 185
C = C (in alkenes)	115 – 140
C ≡ C (in alkynes)	70 – 96
– CH	25 – 50
– CH_2	16 – 40
– CH_3	5 – 15

    In a carbon-13 NMR spectrum, the number of lines correspond to the number of chemically different carbon atoms and the position of the line (the value of the chemical shift) indicates the type of carbon atom.

    The spectrum for propanal is shown.

    Spectrum 1

    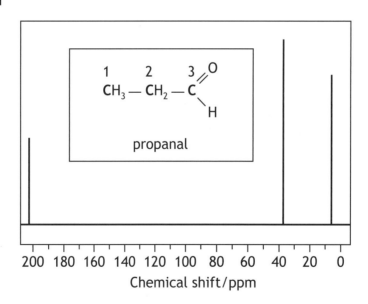

    (i) Use the table of chemical shifts to label each of the peaks on the spectrum with a number to match the carbon atom in propanal that is responsible for the peak.

14. (continued)

(ii) Hydrocarbon **X** has a relative formula mass of 54. Hydrocarbon **X** reacts with hydrogen. One of the products, hydrocarbon **Y**, has a relative formula mass of 56.

The carbon-13 NMR spectrum for hydrocarbon **Y** is shown below.

Spectrum 2

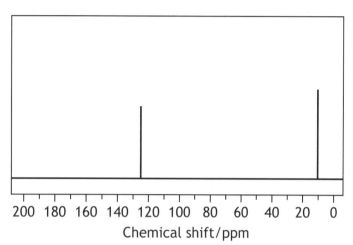

Name hydrocarbon **Y**. **1**

Total marks **2**

[END OF MODEL PAPER]

ADDITIONAL SPACE FOR ANSWERS AND ROUGH WORK

HIGHER FOR CfE
Model Paper 3

Whilst this Model Paper has been specially commissioned by Hodder Gibson for use as practice for the Higher (for Curriculum for Excellence) exams, the key reference documents remain the SQA Specimen Paper 2014 and SQA Past Paper 2015.

National Qualifications
MODEL PAPER 3

Chemistry
Section 1 — Questions

Duration — 2 hours and 30 minutes

Reference may be made to the Chemistry Higher and Advanced Higher Data Booklet.

Instructions for the completion of Section 1 are given on *Page two* of your question and answer booklet.

Record your answers on the answer grid on *Page three* of your question and answer booklet.

Before leaving the examination room you must give your question and answer booklet to the Invigilator; if you do not you may lose all the marks for this paper.

SECTION 1 — 20 marks
Attempt ALL questions

1. The graph shows how the rate of reaction varies with temperature.

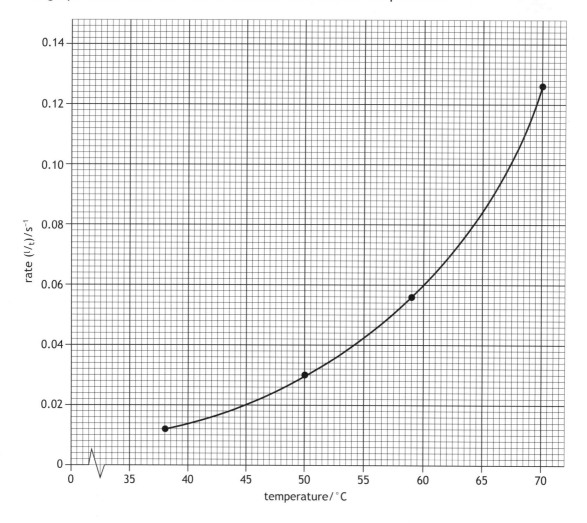

When the temperature was 45°C the reaction time, in seconds, was

A 0.02

B 2250

C 50

D 1.11.

2. The energy profile for a reaction is shown below.

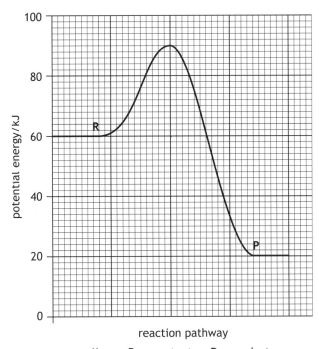

Key: R = reactants P = products

Which letter correctly describes the activation energy and enthalpy change for the forward reaction?

	Activation energy, kJmol^{-1}	Enthalpy change, kJmol^{-1}
A	70	30
B	40	30
C	30	−70
D	30	−40

3. Mg(g) ⟶ Mg^{2+}(g) + 2e$^-$

The energy change, in kJmol^{-1}, for this reaction is

A + 1451

B + 2189

C −2189

D −1451.

Page three

4. Which of the following molecules contains the most polar bond?

    A PF_3
    B CH_3OH
    C H_2S
    D NH_3

5. Which of the following is an example of a polar molecule?

    A CO_2
    B CF_4
    C $CaCl_2$
    D H_2O

6. Which of the following is the best description for the bonding types found in water?

    A   hydrogen and non-polar covalent bonds
    B   hydrogen and polar covalent bonds
    C   London dispersion forces and metallic bonds
    D   London dispersion forces and non-polar covalent bonds

7. Which of the following compounds does **not** contain hydrogen bonding?

    A CH_3OH
    B   NaOH
    C   HF
    D CH_3NH_2

8. Which statement correctly describes the reaction between methanol and ethanoic acid?

	Type of reaction	Product formed
A	Condensation	Ethyl methanoate
B	Hydration	Ethyl methanoate
C	Condensation	Methyl ethanoate
D	Hydration	Methyl ethanoate

9. Which of the following compounds would oxidise to form a single compound which is **not** acidic?

   A ethanal
   B propan-2-ol
   C ethanol
   D propanone

10. The structure of limonene is shown below.

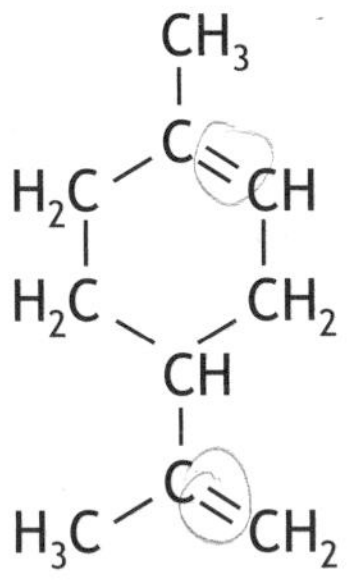

    Which of the following compounds will be the best solvent for limonene?

    A Hexane
    B Hexan-1-ol
    C Hexanal
    D Hexanoic acid

11. Ethanol can react with hot copper(II) oxide to form an aldehyde. Which statement can be applied to this reaction?

    A The Cu^{2+} ion is oxidised.
    B The aldehyde formed contains a hydroxyl group.
    C There is an increase in the oxygen to hydrogen ratio.
    D The ethanol is reduced by the copper(II) oxide.

12. Capsaicin is a flavour molecule found in peppers.

    Which of the following statements **cannot** be applied to capsaicin?

    A   It can be hydrolysed to form a carboxylic acid.

    B   It has a low solubility in water due to the long hydrocarbon chain.

    C   It contains an amide link.

    D   It contains an ester link.

13. Vitamin E is commonly added to skin creams as it can react with damaging free radicals to form stable molecules.

    In this example, Vitamin E is acting as

    A   an initiator of free radical reactions
    B   a free radical scavenger
    C   an oxidising agent
    D   a catalyst.

14. $H_2(g) + Cl_2(g) \rightarrow 2HCl(g)$

    The volume of hydrogen required to produce 71 litres of HCl(g) is

    A   2 litres
    B   4 litres
    C   35.5 litres
    D   71 litres.

Page six

15. 320g of methane was burned and released 17,000kJ of energy. Using this information, the enthalpy of combustion of methane is

    A   −53kJ
    B   −53kJ mol^{-1}
    C   −850kJ
    D   −850kJmol^{-1}.

16. Which statement correctly describes the effect of a catalyst when added to a reversible reaction?

	Position of equilibrium	Time to reach equilibrium
A	Shifts to the right	Equilibrium reached in a shorter time
B	Shifts to the right	Equilibrium reached in a longer time
C	No change	No change
D	No change	Equilibrium reached in a shorter time

17. $N_2(g) + 3H_2(g) \rightleftharpoons 2NH_3(g)$ $\Delta H = -52\,kJ\,mol^{-1}$

    Production of NH_3 will be favoured by

    A   high pressure and high temperature
    B   high pressure and low temperature
    C   low pressure and high temperature
    D   low pressure and low temperature.

18. $H_2(g) + F_2(g) \rightarrow 2HF(g)$

    The enthalpy change for this reaction is equal to

    A   2×(bond enthalpy of H-F) − (bond enthalpy of H-H) + (bond enthalpy of F-F)
    B   2×(bond enthalpy of H-F) − (bond enthalpy of H-H) − (bond enthalpy of F-F)
    C   (bond enthalpy of H-H) − (bond enthalpy of F-F) − 2×(bond enthalpy of H-F)
    D   (bond enthalpy of H-H) + (bond enthalpy of F-F) − 2×(bond enthalpy of H-F).

19. $3CO(g) + Fe_2O_3(s) \rightarrow 3CO_2(g) + 2Fe(s)$

Which statement can be applied to this reaction?

A The Fe^{2+} is reduced

B The Fe^{3+} is oxidised

C CO is acting as a reducing agent

D CO_2 is acting as a reducing agent

20. $Mg(s) + 2HCl(aq) \rightarrow MgCl_2(aq) + H_2(g)$

Magnesium was reacted with hydrochloric acid in a flask with a side-arm. The rate of this reaction could be determined using a timer and

A a thermometer

B a gas syringe

C an upturned test tube in water

D a burette and pipette.

[END OF SECTION 1. NOW ATTEMPT THE QUESTIONS IN SECTION 2 OF YOUR QUESTION AND ANSWER BOOKLET.]

National Qualifications
MODEL PAPER 3

Chemistry
Section 1 — Answer Grid and Section 2

Duration — 2 hours and 30 minutes

Reference may be made to the Chemistry Higher and Advanced Higher Data Booklet.

Total marks — 100

SECTION 1 — 20 marks

Attempt ALL questions.

Instructions for completion of Section 1 are given on *Page 146*.

SECTION 2 — 80 marks

Attempt ALL questions

Write your answers clearly in the spaces provided in this booklet. Additional space for answers and rough work is provided at the end of this booklet. If you use this space you must clearly identify the question number you are attempting. Any rough work must be written in this booklet. You should score through your rough work when you have written your final copy.

Use **blue** or **black** ink.

Before leaving the examination room you must give this booklet to the Invigilator; if you do not you may lose all the marks for this paper.

SECTION 1—20 marks

The questions for Section 1 are contained on *Page 137* — Questions.
Read these and record your answers on the answer grid on *Page 147* opposite.
DO NOT use gel pens.

1. The answer to each question is **either** A, B, C or D. Decide what your answer is, then fill in the appropriate bubble (see sample question below).

2. There is **only one correct** answer to each question.

3. Any rough working should be done on the additional space for answers and rough work at the end of this booklet.

Sample Question

To show that the ink in a ball-pen consists of a mixture of dyes, the method of separation would be:

    A   fractional distillation

    B   chromatography

    C   fractional crystallisation

    D   filtration.

The correct answer is **B**—chromatography. The answer **B** bubble has been clearly filled in (see below).

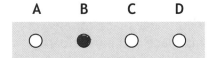

Changing an answer

If you decide to change your answer, cancel your first answer by putting a cross through it (see below) and fill in the answer you want. The answer below has been changed to **D**.

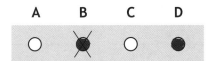

If you then decide to change back to an answer you have already scored out, put a tick (✓) to the **right** of the answer you want, as shown below:

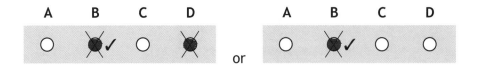

SECTION 1 — Answer Grid

	A	B	C	D
1	○	○	○	○
2	○	○	○	○
3	○	○	○	○
4	○	○	○	○
5	○	○	○	○
6	○	○	○	○
7	○	○	○	○
8	○	○	○	○
9	○	○	○	○
10	○	○	○	○
11	○	○	○	○
12	○	○	○	○
13	○	○	○	○
14	○	○	○	○
15	○	○	○	○
16	○	○	○	○
17	○	○	○	○
18	○	○	○	○
19	○	○	○	○
20	○	○	○	○

SECTION 2 — 80 marks

Attempt ALL questions

1. Information about four elements from the third period of the Periodic Table is shown in the table.

Element	aluminium	silicon	phosphorus	sulfur
Bonding		covalent		covalent
Structure	lattice		molecular	

   (a) Complete the table to show the bonding and structure for each element. **2**

   (b) Why is there a decrease in the size of atoms across the period from aluminium to sulfur? **1**

   (c) Argon is also in the third period. Argon is a very useful gas and each year 750 000 tonnes of argon are extracted from liquid air.

   (i) Suggest how argon could be extracted from liquid air. **1**

   (ii) Air contains 1·3% argon by mass. Calculate the mass of liquid air needed to obtain 750 000 tonnes of argon. **1**

   (iii) Argon is used in the manufacture of magnesium powder. A jet of liquid argon is blown at a stream of molten magnesium producing fine droplets of metal. These cool to form the powder.

   Why can liquid air not be used to make magnesium powder? **1**

1. (c) (continued)

   (iv) Argon was discovered in 1890's when samples of nitrogen prepared by different methods were compared. The element name was derived from the Greek *argos*, which means "lazy one".

   Two samples of nitrogen can be prepared as shown.

   Method 1   Removing carbron dioxide and oxygen from the air.

   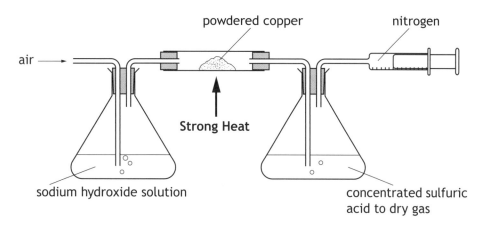

   Method 2   Reaction of sodium nitrite with ammonium chloride.

   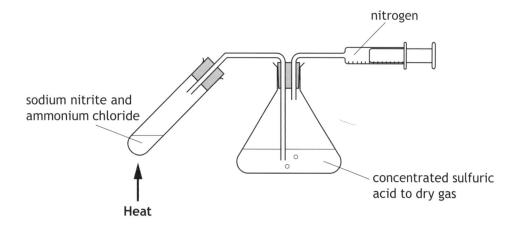

   $$NaNO_2(s) + NH_4Cl(s) \rightarrow Na4Cl(s) + N_2(g) + 2H_2O(\ell)$$

   Heated magnesium metal can react with nitrogen gas to give magnesium nitride.

   $$3Mg(s) + N_2(g) \rightarrow Mg_3N_2(s)$$

MARKS | DO NOT WRITE IN THIS MARGIN

1. (c) **(continued)**

Using your knowledge of chemistry, comment on the discovery and naming of argon. **3**

Total marks 9

2. (a) Hydrogen and chlorine gases are used in an experiment to demonstrate a free radical reaction.

A plastic bottle is wrapped with black tape leaving a "window" on one side. The bottle is filled with a mixture of hydrogen and chlorine. When bright light shines on the bottle there is an explosion.

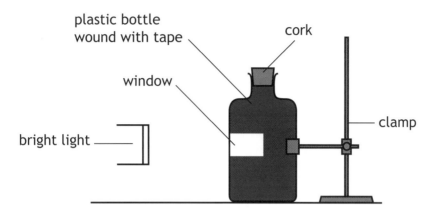

A free radical chain reaction is initiated when light energy causes chlorine radicals to form as shown below.

Initialisation $Cl-Cl \xrightarrow{light} Cl\bullet + Cl\bullet$

(i) Complete the equations below showing possible propagation and termination steps.

Propagation $Cl\bullet + H-H \longrightarrow Cl + H\bullet$

Termination $H\bullet + H\bullet \longrightarrow H_2$

(ii) Why is the plastic bottle used in the experiment wrapped in black tape?

2. (continued)

(b) The production of hydrogen chloride from hydrogen and chlorine is exothermic.

$$H_2(g) + Cl_2(g) \rightarrow 2HCl(g)$$

Using bond enthalpy values, calculate the enthalpy change, in kJ mol^{-1}, for this reaction.

Total marks 10

3. When vegetable oils are hydrolysed, mixtures of fatty acids are obtained. The fatty acids can be classified by their degree of unsaturation.

The table below shows the composition of each of the mixtures of fatty acids obtained when palm oil and olive oil were hydrolysed.

	Palm oil	Olive oil
Saturated fatty acids	51%	16%
Monounsaturated fatty acids	39%	75%
Polyunsaturated fatty acids	10%	9%

(a) Why does palm oil have a higher melting point than olive oil?

MARKS | DO NOT WRITE IN THIS MARGIN

3. (continued)

(b) One of the fatty acids produced by the hydrolysis of palm oil is linoleic acid, $C_{17}H_{31}COOH$.

To which class (saturated, monounsaturated or polyunsaturated) does this fatty acid belong? **1**

(c) When a mixture of palm oil and olive oil is hydrolysed using a solution of sodium hydroxide, a mixture of the sodium salts of the fatty acids is obtained.

(i) State a use for these fatty acid salts. **1**

(ii) name the other product of the hydrolysis reaction. **1**

(d) Over time, open containers of oils develop a rancid flavour. What substance is reacting with the oil to cause these unwanted changes to take place? **1**

Total marks 5

4. Zinc can be used to coat iron objects as it prevents them from rusting.

A student calculated the mass of zinc on some iron nails by reacting the nails with hydrochloric acid. He measured the volume of hydrogen gas produced using an upturned measuring cylinder, as shown below.

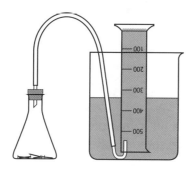

(a) Suggest an alternative method for measuring the volume of hydrogen gas. **1**

(b) Another student suggested heating the nails and acid mixture using a Bunsen burner to speed up the rate of the reaction.

   (i) Why does increasing the temperature increase the rate of a chemical reaction? **1**

   (ii) Why is a Bunsen burner not suitable for heating in this experiment? **1**
   It is flammable.

(c) Zinc is also an essential element for the body and is found in a variety of foods.

A label from a packet of peanuts stated that 100g of peanuts would provide 3.3mg of zinc which is equivalent to 22% of the recommended daily allowance.

Calculate the mass of peanuts, in grams, required to provide the recommended daily allowance of zinc. **2**

Total marks **5**

5. Alcohols can be oxidised to carboxylic acids.

$CH_3CH_2CH_2OH$ **Step 1** → CH_3CH_2CHO **Step 2** → CH_3CH_2COOH

propan-1-ol   propanal   propanoic acid

(a) Why can Step 1 be described as an oxidation reaction? **1**

(b) Propan-1-ol and propanoic acid react to form an ester.

The mixture of excess reactants and ester product is poured onto sodium hydrogencarbonate solution.

(i) What evidence would show that an ester is formed? **1**

(ii) Draw a structural formula for this ester. **1**

Total marks 3

6. Phenylalanine and alanine are both amino acids.

phenylalanine   alanine

(a) Phenylalanine is an essential amino acid. What is meant by an essential amino acid? **1**

(b) Phenylalanine and alanine can react to form the dipeptide shown.

Circle the peptide link in this molecule. **1**

(c) Draw a structural formula for the other dipeptide that can be formed from phenylalanine and alanine. **1**

Total marks **3**

7. Whisky contains many flavour compounds such as aldehydes, esters and alcohols.

Techniques for measuring the alcohol strength and measuring the concentration of the flavour compounds in whisky are routinely used to determine whether a whisky is genuine or fake.

(a) The concentration of flavour compounds can be measured by Gas Chromatography. The gas chromatogram from a sample of genuine whisky is shown below.

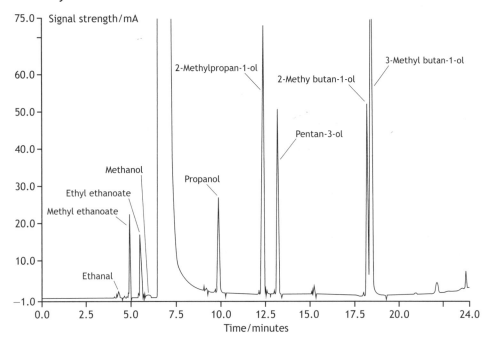

(i) Explain how this chromatogram could be used to determine if a whisky is genuine or fake. **1**

(ii) Why does methanol have a longer retention time than ethanal? **1**

(iii) Predict the retention time for ethanol. **1**

(iv) Why is it difficult to obtain accurate values for the amount of 2-methylbutan-1-ol present in a sample containing large amounts of 3-methylbutan-1-ol? **1**

MARKS | DO NOT WRITE IN THIS MARGIN

7. (continued)

(b) Furaneol and vanillin are two flavour compounds found in whisky.

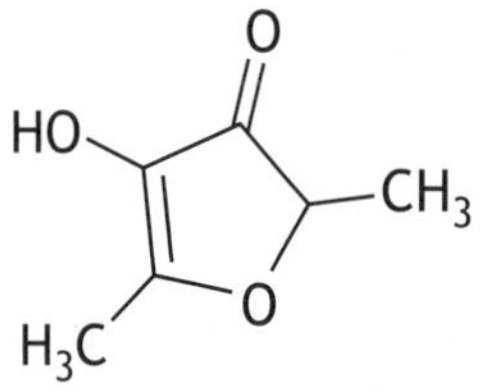

Furaneol

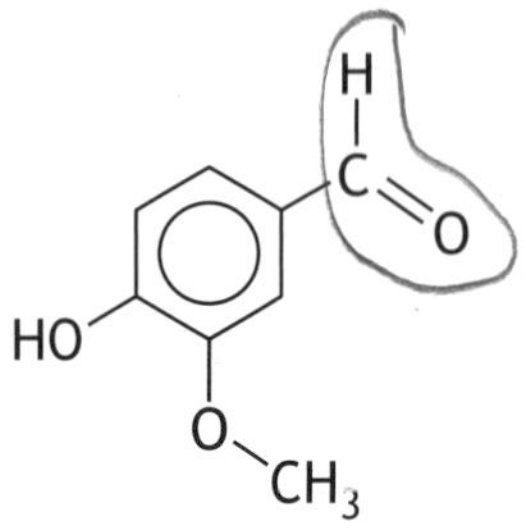

Vanillin

(i) Vanillin is an example of an aldehyde. Circle the part of the Vanillin molecule that allows you to state that Vanillin is an aldehyde. **1**

(ii) Name a chemical that could be used to distinguish between furaneol and vanillin. Tollen's Reagant **1**

(iii) Both flavour compounds are soluble in water as they can hydrogen bond to water.

In the box below showing a molecule of furaneol, draw a molecule of water and use a dotted line to show where a hydrogen bond can exist between the two molecules. **1**

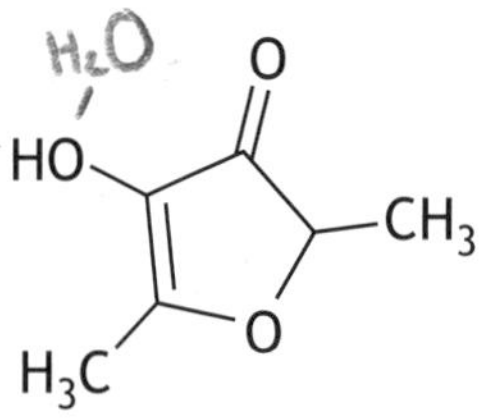

Furaneol

MARKS | DO NOT WRITE IN THIS MARGIN

7. (continued)

(c) The flavour of 100 year old whisky is different from the flavour of freshly made whisky. **Using your knowledge of chemistry**, describe some of the changes that a whisky could undergo and how this could change the flavour. In your answer you should refer to the names or structures of compounds and describe how they change. **3**

Total marks 10

MARKS | DO NOT WRITE IN THIS MARGIN

8. In the lab, nitrogen dioxide gas can be prepared by heating copper(II) nitrate.

$$Cu(NO_3)_2(s) \rightarrow CuO(s) + 2NO_2(g) + \tfrac{1}{2}O_2(g)$$

(a) Calculate the volume, in litres, of nitrogen dioxide gas produced when 2.0 g of copper(II) nitrate is completely decomposed on heating. **3**

(Take the molar volume of nitrogen dioxide to be 24 litres mol^{-1}.)

Show your working clearly.

(b) Nitrogen dioxide has a boiling point of 22 °C.

Complete the diagram to show how nitrogen dioxide can be separated and collected. **1**

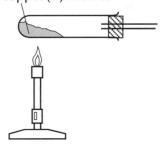

copper(II) nitrate
oxygen gas
water

Total marks 4

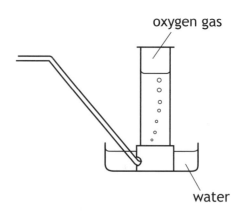

9. In many bathroom cleaning products, the bleaching agent is the hypochlorite ion, ClO⁻(aq).

   (a) Hypochlorite bleaches can be made by reacting sodium hydroxide with chlorine. Sodium hypochlorite, sodium chloride and water are formed.

   Write a balanced equation for the reaction. **2**

   (b) In the bleach solution, the following equilibrium exists.

   $$Cl_2(aq) + H_2O(\ell) \rightleftharpoons 2H^+(aq) + ClO^-(aq) + Cl^-(aq)$$

   Explain why the addition of sodium hydroxide increases the bleaching efficiency of the solution. **2**

   (c) When ClO⁻(aq) acts as a bleach, it is reduced to produce the ClO⁻(aq) ion.

   $$ClO^-(aq) \rightarrow Cl^-(aq)$$

   Complete the above to form the ion-electron equation for the reduction reaction. **1**

9. (continued)

(d) An experiment was carried out to measure the concentration of hypochlorite ions in a sample of bleach. In this experiment, the bleach sample reacted with excess hydrogen peroxide.

$$H_2O_2(aq) + ClO^-(aq) \rightarrow H_2O(\ell) + Cl^-(aq) + O_2(g)$$

By measuring the volume of oxygen given off, the concentration of bleach can be calculated.

80 cm³ of oxygen gas was produced from 5·0 cm³ of bleach.

Calculate the concentration of the hypochlorite ions in the bleach.

(Take the molecular volume of oxygen to be 24 litre mol⁻¹.) 4

10. Hairspray is a mixture of chemicals.

(a) A primary alcohol, 2-methylpropan-1-ol, is added to hairspray to help it dry quickly on the hair.

$$H-\overset{\overset{\displaystyle H}{|}}{\underset{\underset{\displaystyle H}{|}}{C}}-\overset{\overset{\displaystyle CH_3}{|}}{\underset{\underset{\displaystyle H}{|}}{C}}-\overset{\overset{\displaystyle H}{|}}{\underset{\underset{\displaystyle H}{|}}{C}}-OH$$

Draw a structural formula for a secondary alcohol that is an isomer of 2-methylpropan-1-ol.

1

MARKS DO NOT WRITE IN THIS MARGIN

10. (continued)

(b) Triethanol amine and triisopropyl amine are bases used to neutralise acidic compounds in the hairspray to prevent damage to the hair.

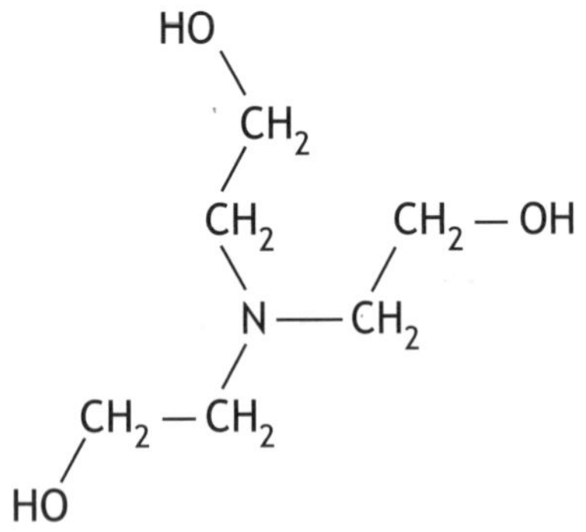

triethanol amine

molecular mass 149
boiling point 335 °C

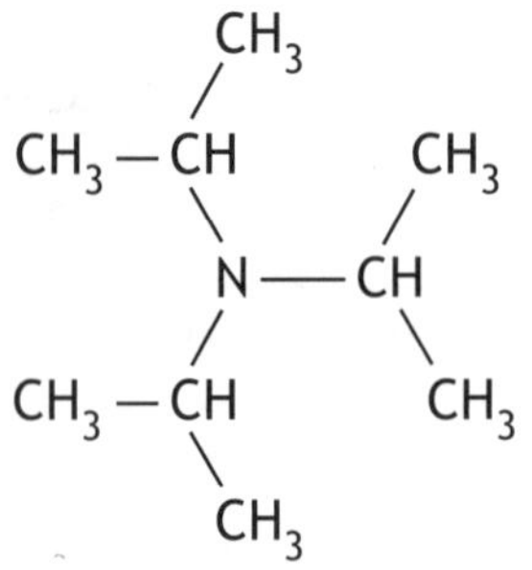

triisopropyl amine

molecular mass 143
boiling point 47 °C

In terms of the intermolecular bonding present, **explain clearly** why triethanol amine has a much higher boiling point than triisopropyl amine. In your answer you should mention the intermolecular forces involved and how they arise. **3**

Total marks 4

11. Aspirin, a common pain-killer, can be made by the reaction of salicylic acid with ethanoic anhydride.

$C_7H_6O_3$	$C_4H_6O_3$	$C_9H_8O_4$	$C_2H_4O_2$
salicylic acid	ethanoic anhydride	aspirin	ethanoic acid
mass of one mole = 138 g	mass of one mole = 102 g	mass of one mole = 180 g	mass of one mole = 60 g

(a) Calculate the atom economy for the formation of aspirin using this method.

Show your working clearly. 2

(b) In a laboratory preparation of aspirin, 5·02 g of salicylic acid produced 2·62 g of aspirin.

Calculate the percentage yield of aspirin. 2

Show your working clearly.

Total marks 4

12. Different fuels are used for different purposes.

    (a) Ethanol, C_2H_5OH, can be used as a fuel in some camping stoves.

    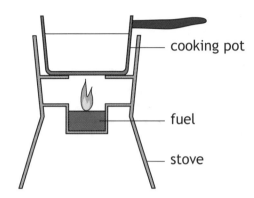

    (i) The enthalpy of combustion of ethanol given in the data booklet is -1367 kJ mol^{-1}.

    Using this value, calculate the mass of ethanol, in g, required to raise the temperature of 500 g of water from 18°C to 100°C.

    Show your working clearly.

    (ii) Suggest a reason why less energy is obtained from burning ethanol in the camping stove than is predicted from its enthalpy of combustion.

MARKS | DO NOT WRITE IN THIS MARGIN

12. (continued)

(b) Petrol is a fuel used in cars.

Energy released when 1·00 g of petrol burned/kJ	48·0
Volume of 1·00 g of petrol/cm^3	1·45

A car has a 50·0 litre petrol tank.

Calculate the energy, in kJ, released by the complete combustion of one tank of petrol. **2**

Total marks 6

13. The element boron forms many useful compounds.

(a) Borane (BH₃) is used to synthesise alcohols from alkenes.

The reaction occurs in two stages

Stage 1 Addition Reaction

The boron atom bonds to the carbon atom of the double bond which already has the most hydrogens **directly** attached to it.

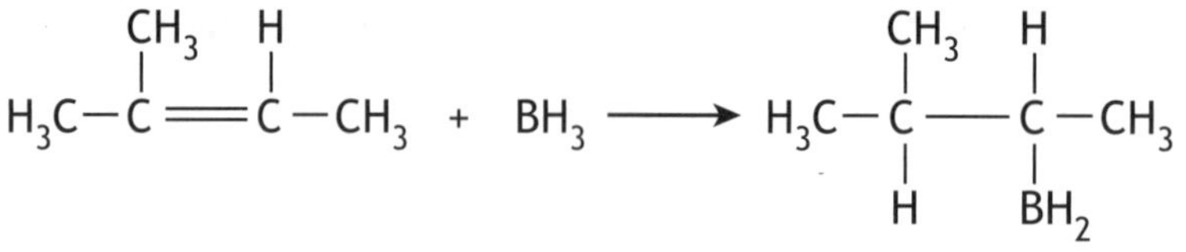

$$H_3C-\underset{\underset{CH_3}{|}}{C}=\underset{\underset{H}{|}}{C}-CH_3 + BH_3 \longrightarrow H_3C-\underset{\underset{H}{|}}{\overset{\overset{CH_3}{|}}{C}}-\underset{\underset{BH_2}{|}}{\overset{\overset{H}{|}}{C}}-CH_3$$

Stage 2 Oxidation Reaction

The organoborane compound is oxidised to form the alcohol.

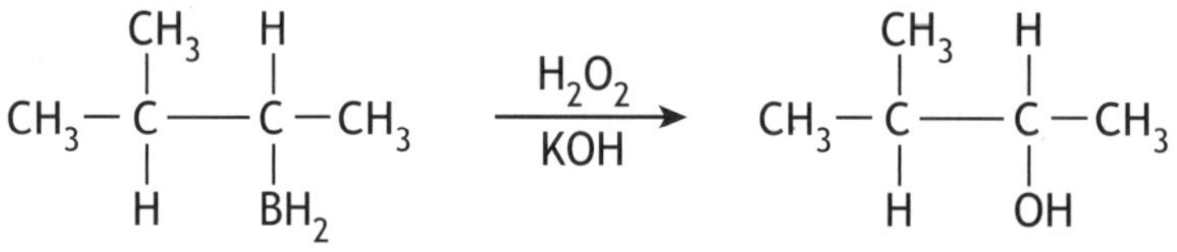

$$CH_3-\underset{\underset{H}{|}}{\overset{\overset{CH_3}{|}}{C}}-\underset{\underset{BH_2}{|}}{\overset{\overset{H}{|}}{C}}-CH_3 \xrightarrow[KOH]{H_2O_2} CH_3-\underset{\underset{H}{|}}{\overset{\overset{CH_3}{|}}{C}}-\underset{\underset{OH}{|}}{\overset{\overset{H}{|}}{C}}-CH_3$$

(i) Name the alcohol produced in Stage 2. **1**

(ii) Draw a structural formula for the alcohol which would be formed from the alkene shown below. **1**

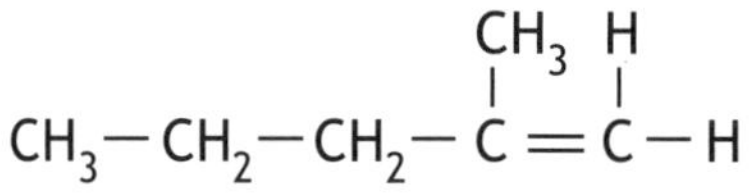

$$CH_3-CH_2-CH_2-\underset{\underset{H}{|}}{\overset{\overset{CH_3}{|}}{C}}=\underset{\underset{H}{|}}{C}-H$$

MARKS | DO NOT WRITE IN THIS MARGIN

13. (continued)

(b) The equation for the combustion of diborane is shown below.

$$B_2H_6(g) + 3O_2(g) \rightarrow B_2O_3(s) + 3H_2O(\ell)$$

Calculate the enthalpy of combustion of diborane (B_2H_6) in kJ mol^{-1}, using the following data. **2**

$$2B(s) + 3H_2(g) \rightarrow B_2H_6(g) \quad \Delta H = -36 \text{ kJ mol}^{-1}$$

$$H_2(g) + ½O_2(g) \rightarrow H_2O(\ell) \quad \Delta H = -286 \text{ kJ mol}^{-1}$$

$$2B(s) + 1½O_2(g) \rightarrow B_2O_3(s) \quad \Delta H = -1274 \text{ kJ mol}^{-1}$$

Total marks 5

MARKS | DO NOT WRITE IN THIS MARGIN

14. Solutions containing iodine are used to treat foot rot in sheep.

The concentration of iodine in a solution can be determined by titrating with a solution of thiosulfate ions.

$$I_2 \quad + \quad 2S_2O_3^{2-} \quad \rightarrow \quad 2I^- \quad + \quad S_4O_6^{2-}$$

thiosulfate ions

Three 20·0 cm^3 samples of a sheep treatment solution were titrated with 0·10 mol l^{-1} thiosulfate solution.

The results are shown below.

Sample	Volume of thiosulfate/cm^3
1	18·60
2	18·10
3	18·20

(i) Why is the volume of sodium thiosulfate used in the calculation taken to be 18·15 cm^3, although this is not the average of the three titres in the table? **1**

MARKS DO NOT WRITE IN THIS MARGIN

14. (continued)

(ii) Calculate the concentration of iodine, in mol l^{-1}, in the foot rot treatment solution. 3

Show your working clearly.

(iii) Describe how to prepare 250 cm^3 of a 0·10 mol l^{-1} standard solution of sodium thiosulfate, $Na_2S_2O_3$. 3

Your answer should include the mass, in g, of sodium thiosulfate required.

Total marks 7

15. The boiling point of water can be raised by the addition of a solute.

The increase in boiling point depends only on the **number** of solute particles but not the type of particle.

The increase in boiling point (ΔT_b), in °C, can be estimated using the formula shown.

$$\Delta T_b = 0.51 \times c \times i$$

where

  c is the concentration of the solution in mol l^{-1}.

  i is the number of particles released into solution when one formula unit of the solute dissolves.

The value of i for a number of compounds is shown in the table below.

Solute	i
NaCl	2
MgCl$_2$	3
(NH$_4$)$_3$PO$_4$	4

(a) What is the value of i for sodium sulphate? **1**

(b) Calculate the increase in boiling point, ΔT_b, for a 0·10 mol l^{-1} solution of ammonium phosphate. **1**

[END OF MODEL PAPER]

ADDITIONAL SPACE FOR ANSWERS AND ROUGH WORK

ADDITIONAL SPACE FOR ANSWERS AND ROUGH WORK

HIGHER FOR CfE
2015

National Qualifications 2015

X713/76/02

**Chemistry
Section 1 — Questions**

THURSDAY, 28 MAY
1:00 PM – 3:30 PM

Instructions for the completion of Section 1 are given on *Page two* of your question and answer booklet X713/76/01.

Record your answers on the answer grid on *Page three* of your question and answer booklet.

Reference may be made to the Chemistry Higher and Advanced Higher Data Booklet.

Before leaving the examination room you must give your question and answer booklet to the Invigilator; if you do not you may lose all the marks for this paper.

SECTION 1 — 20 marks
Attempt ALL questions

1. The elements nitrogen, oxygen, fluorine and neon

   A can form negative ions

   B are made up of diatomic molecules

   C have single bonds between the atoms

   D are gases at room temperature.

2. Which of the following equations represents the first ionisation energy of fluorine?

   A $F^-(g) \rightarrow F(g) + e^-$

   B $F^-(g) \rightarrow \frac{1}{2}F_2(g) + e^-$

   C $F(g) \rightarrow F^+(g) + e^-$

   D $\frac{1}{2}F_2(g) \rightarrow F^+(g) + e^-$

3. Which of the following atoms has least attraction for bonding electrons?

   A Carbon

   B Nitrogen

   C Phosphorus

   D Silicon

4. Which of the following is **not** an example of a van der Waals' force?

   A Covalent bond

   B Hydrogen bond

   C London dispersion force

   D Permanent dipole - permanent dipole attraction

5. Which of the following has more than one type of van der Waals' force operating between its molecules in the liquid state?

   A   Br—Br

   B   O=C=O

   C   H—N(H)(H)   (ammonia)

   D   H—C(H)(H)—H   (methane)

6. Oil molecules are more likely to react with oxygen in the air than fat molecules.

   During the reaction the oil molecules

   A   are reduced
   B   become rancid
   C   are hydrolysed
   D   become unsaturated.

7. Which of the following mixtures will form when NaOH(aq) is added to a mixture of propanol and ethanoic acid?

   A   Propanol and sodium ethanoate
   B   Ethanoic acid and sodium propanoate
   C   Sodium hydroxide and propyl ethanoate
   D   Sodium hydroxide and ethyl propanoate

8. Oils contain carbon to carbon double bonds which can undergo addition reactions with iodine.

   The iodine number of an oil is the mass of iodine in grams that will react with 100 g of oil.

   Which line in the table shows the oil that is likely to have the lowest melting point?

	Oil	Iodine number
A	Corn	123
B	Linseed	179
C	Olive	81
D	Soya	130

9. When an oil is hydrolysed, which of the following molecules is always produced?

A
COOH
|
CHOH
|
COOH

B
CH$_2$OH
|
CHOH
|
CH$_2$OH

C C$_{17}$H$_{35}$COOH

D C$_{17}$H$_{33}$COOH

10. Enzymes are involved in the browning of cut fruit.

One reaction taking place is:

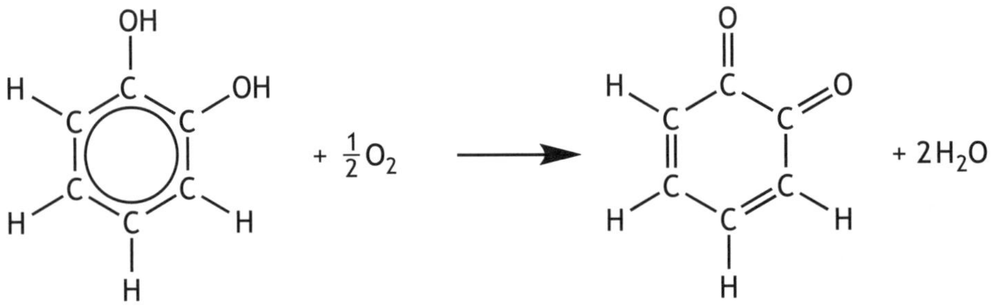

Which of the following correctly describes the above reaction?

A Oxidation
B Reduction
C Hydrolysis
D Condensation

11. Which of the following statements is correct for ketones?

A They are formed by oxidation of tertiary alcohols.

B They contain the group —CHO.

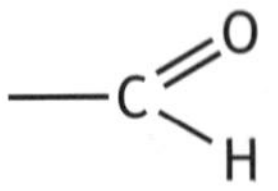

C They contain a carboxyl group.

D They will not react with Fehling's solution.

12. Carvone is a natural product that can be extracted from orange peel.

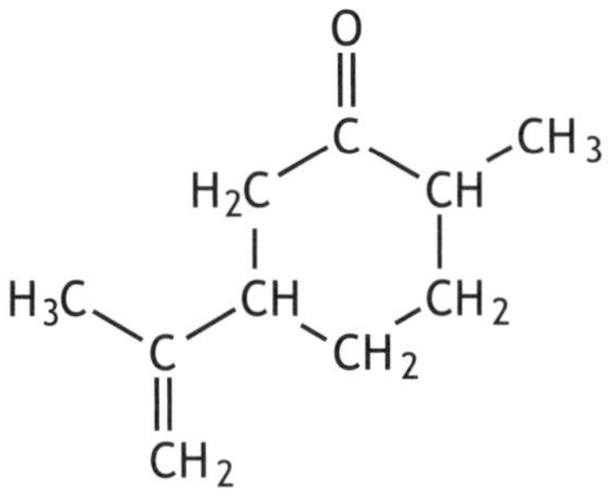

Carvone

Which line in the table correctly describes the reaction of carvone with bromine solution and with acidified potassium dichromate solution?

	Reaction with bromine solution	Reaction with acidified potassium dichromate solution
A	no reaction	no reaction
B	no reaction	orange to green
C	decolourises	orange to green
D	decolourises	no reaction

13. The structure of isoprene is

A

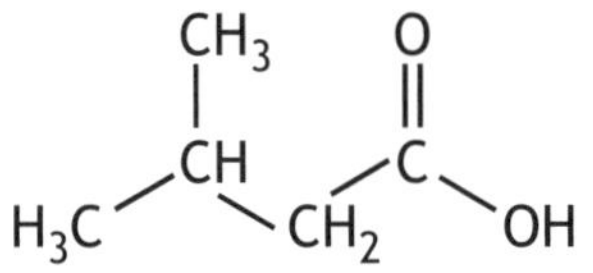

B

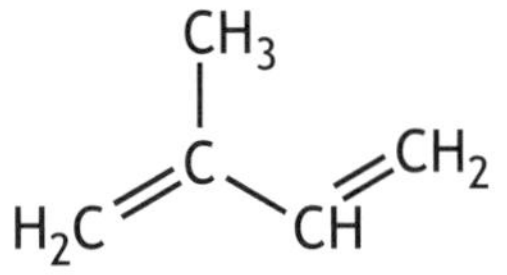

C

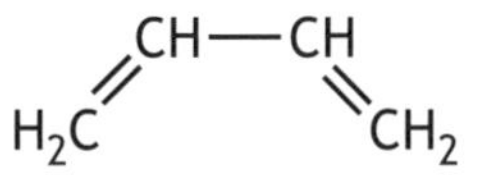

D

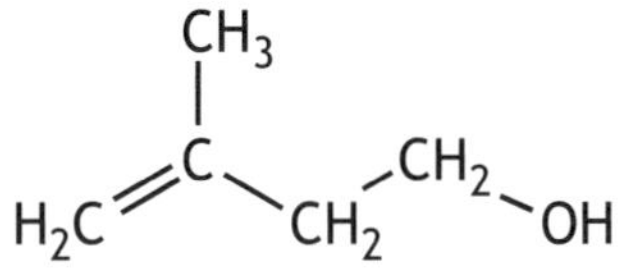

[Turn over

14. The antibiotic, erythromycin, has the following structure.

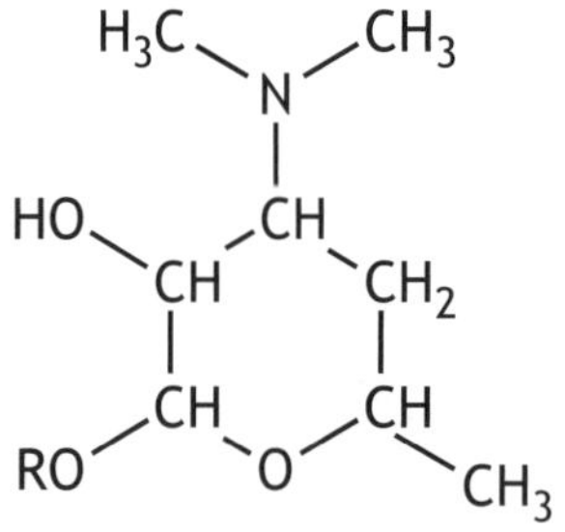

To remove its bitter taste, the erythromycin is reacted to give the compound with the structure shown below.

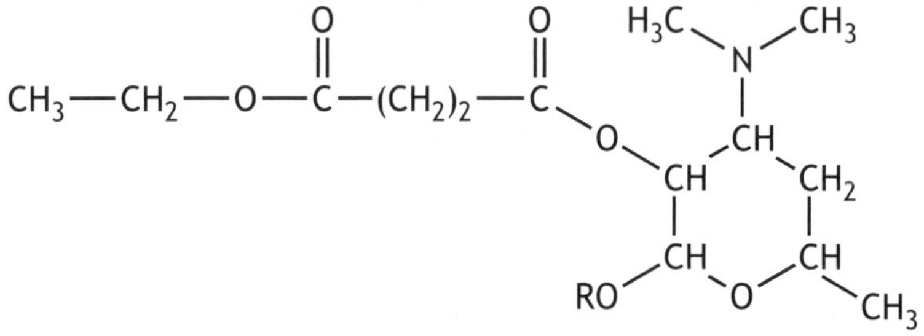

Which of the following types of compound has been reacted with erythromycin to produce this compound?

A Alcohol
B Aldehyde
C Carboxylic acid
D Ketone

15. Which of the following is an isomer of 2,2-dimethylpentan-1-ol?

A $CH_3CH_2CH_2CH(CH_3)CH_2OH$
B $(CH_3)_3CCH(CH_3)CH_2OH$
C $CH_3CH_2CH_2CH_2CH_2CH_2CH_2CH_2OH$
D $(CH_3)_2CHC(CH_3)_2CH_2CH_2OH$

16. Consider the reaction pathway shown below.

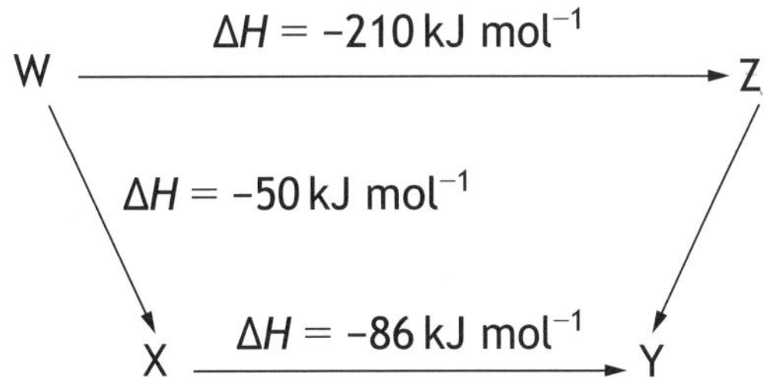

$$W \xrightarrow{\Delta H = -210 \text{ kJ mol}^{-1}} Z$$
$$W \xrightarrow{\Delta H = -50 \text{ kJ mol}^{-1}} X$$
$$X \xrightarrow{\Delta H = -86 \text{ kJ mol}^{-1}} Y$$

According to Hess's Law, the ΔH value, in kJ mol^{-1}, for reaction Z to Y is

A +74

B −74

C +346

D −346.

[Turn over

17. $I_2(s) \rightarrow I_2(g)$ $\Delta H = +60 \text{ kJ mol}^{-1}$

    $I_2(g) \rightarrow 2I(g)$ $\Delta H = +243 \text{ kJ mol}^{-1}$

    $I(g) + e^- \rightarrow I^-(g)$ $\Delta H = -349 \text{ kJ mol}^{-1}$

Which of the following would show the energy diagram for $I_2(s) + 2e^- \rightarrow 2I^-(g)$?

A
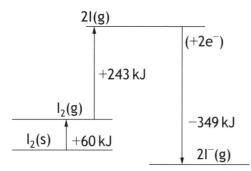

B
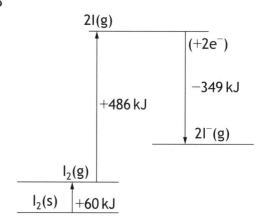

C
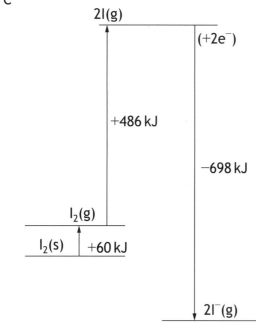

D
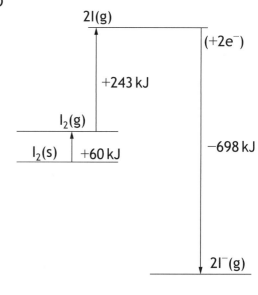

18. Which of the following statements regarding a chemical reaction at equilibrium is always correct?

    A The rates of the forward and reverse reactions are equal.

    B The concentration of reactants and products are equal.

    C The forward and reverse reactions have stopped.

    D The addition of a catalyst changes the position of the equilibrium.

19. A reaction has the following potential energy diagram.

    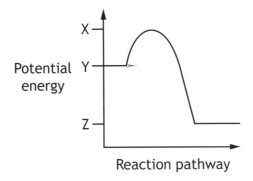

    The activation energy for the forward reaction is

    A X – Y
    B Y – X
    C Y – Z
    D Z – Y.

20. Which of the following will react with Br$_2$ but **not** with I$_2$?

    A OH$^-$
    B SO$_3^{2-}$
    C Fe^{2+}
    D Mn^{2+}

[END OF SECTION 1. NOW ATTEMPT THE QUESTIONS IN SECTION 2 OF YOUR QUESTION AND ANSWER BOOKLET.]

FOR OFFICIAL USE

H

National Qualifications 2015

Mark

X713/76/01

Chemistry
Section 1 — Answer Grid
and Section 2

THURSDAY, 28 MAY
1:00 PM – 3:30 PM

Fill in these boxes and read what is printed below.

Full name of centre

Town

Forename(s)

Surname

Number of seat

Date of birth
Day   Month   Year

Scottish candidate number

Total marks — 100

SECTION 1 — 20 marks

Attempt ALL questions.

Instructions for completion of Section 1 are given on *Page two*.

SECTION 2 — 80 marks

Attempt ALL questions

Reference may be made to the Chemistry Higher and Advanced Higher Data Booklet.

Write your answers clearly in the spaces provided in this booklet. Additional space for answers and rough work is provided at the end of this booklet. If you use this space you must clearly identify the question number you are attempting. Any rough work must be written in this booklet. You should score through your rough work when you have written your final copy.

Use **blue** or **black** ink.

Before leaving the examination room you must give this booklet to the Invigilator; if you do not you may lose all the marks for this paper.

SECTION 1—20 marks

The questions for Section 1 are contained in the question paper X713/76/02.
Read these and record your answers on the answer grid on *Page three* opposite.
Use **blue** or **black** ink. Do NOT use gel pens or pencil.

1. The answer to each question is **either** A, B, C or D. Decide what your answer is, then fill in the appropriate bubble (see sample question below).

2. There is **only one correct** answer to each question.

3. Any rough working should be done on the additional space for answers and rough work at the end of this booklet.

Sample Question

To show that the ink in a ball-pen consists of a mixture of dyes, the method of separation would be:

    A   fractional distillation

    B   chromatography

    C   fractional crystallisation

    D   filtration.

The correct answer is **B**—chromatography. The answer **B** bubble has been clearly filled in (see below).

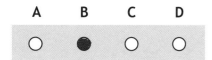

Changing an answer

If you decide to change your answer, cancel your first answer by putting a cross through it (see below) and fill in the answer you want. The answer below has been changed to **D**.

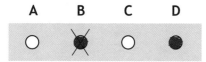

If you then decide to change back to an answer you have already scored out, put a tick (✓) to the **right** of the answer you want, as shown below:

SECTION 1 — Answer Grid

	A	B	C	D
1	○	○	○	○
2	○	○	○	○
3	○	○	○	○
4	○	○	○	○
5	○	○	○	○
6	○	○	○	○
7	○	○	○	○
8	○	○	○	○
9	○	○	○	○
10	○	○	○	○
11	○	○	○	○
12	○	○	○	○
13	○	○	○	○
14	○	○	○	○
15	○	○	○	○
16	○	○	○	○
17	○	○	○	○
18	○	○	○	○
19	○	○	○	○
20	○	○	○	○

SECTION 2 — 80 marks

Attempt ALL questions

1. Volcanoes produce a variety of molten substances, including sulfur and silicon dioxide.

   (a) Complete the table to show the strongest type of attraction that is broken when each substance melts.

Substance	Melting point (°C)	Strongest type of attraction broken when substance melts
sulfur	113	
silicon dioxide	1610	

   (b) Volcanic sulfur can be put to a variety of uses. One such use involves reacting sulfur with phosphorus to make a compound with formula P_4S_3.

   (i) Draw a possible structure for P_4S_3.

   (ii) Explain why the covalent radius of sulfur is smaller than that of phosphorus.

MARKS | DO NOT WRITE IN THIS MARGIN

1. (b) (continued)

(iii) The melting point of sulfur is much higher than that of phosphorus.

Explain fully, in terms of the structures of sulfur and phosphorus molecules and the intermolecular forces between molecules of each element, why the melting point of sulfur is much higher than that of phosphorus. **3**

[Turn over

2. (a) A student investigated the effect of changing acid concentration on reaction rate. Identical strips of magnesium ribbon were dropped into different concentrations of excess hydrochloric acid and the time taken for the magnesium to completely react recorded.

A graph of the student's results is shown below.

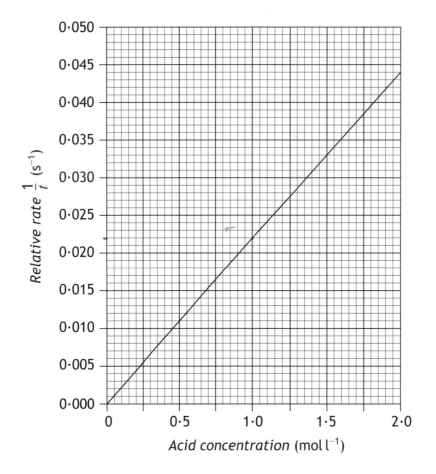

Use information from the graph to calculate the reaction time, in seconds, when the concentration of the acid was 1·0 mol l⁻¹.

2. (continued)

(b) The rate of reaction can also be altered by changing the temperature or using a catalyst.

(i) Graph 1 shows the distribution of kinetic energies of molecules in a gas at 100 °C.

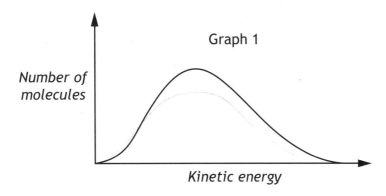

Add a second curve to graph 1 to show the distribution of kinetic energies at 50 °C. **1**

(ii) In graph 2, the shaded area represents the number of molecules with the required activation energy, E_a.

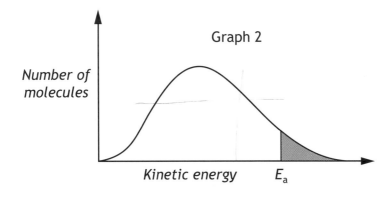

Draw a line to show how a catalyst affects the activation energy. **1**

[Turn over

3. (a) Methyl cinnamate is an ester used to add strawberry flavour to foods. It is a naturally occurring ester found in the essential oil extracted from the leaves of strawberry gum trees.

To extract the essential oil, steam is passed through shredded strawberry gum leaves. The steam and essential oil are then condensed and collected.

(i) Complete the diagram to show an apparatus suitable for carrying out this extraction.

(An additional diagram, if required, can be found on *Page thirty-seven*).

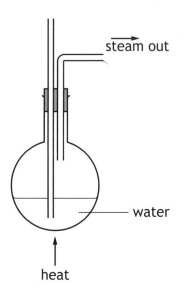

(ii) The essential oil extracted is a mixture of compounds.

Suggest a technique that could be used to separate the mixture into pure compounds.

(b) A student prepared a sample of methyl cinnamate from cinnamic acid and methanol.

cinnamic acid + methanol → methyl cinnamate + water
mass of one mole = 148 g   mass of one mole = 32 g   mass of one mole = 162 g

6·5 g of cinnamic acid was reacted with 2·0 g of methanol.

MARKS | DO NOT WRITE IN THIS MARGIN

3. (b) (continued)

(i) Show, by calculation, that cinnamic acid is the limiting reactant. (One mole of cinnamic acid reacts with one mole of methanol.) **2**

(ii) (A) The student obtained 3·7 g of methyl cinnamate from 6·5 g of cinnamic acid.

Calculate the percentage yield. **2**

(B) The student wanted to scale up the experiment to make 100 g of methyl cinnamate.

Cinnamic acid costs £35·00 per 250 g.

Calculate the cost of cinnamic acid needed to produce 100 g of methyl cinnamate. **2**

[Turn over

4. Up to 10% of perfumes sold in the UK are counterfeit versions of brand name perfumes.

One way to identify if a perfume is counterfeit is to use gas chromatography. Shown below are gas chromatograms from a brand name perfume and two different counterfeit perfumes. Some of the peaks in the brand name perfume have been identified as belonging to particular compounds.

Brand name perfume

Ⓐ linalool
Ⓑ citronellol
Ⓒ geraniol
Ⓓ eugenol
Ⓔ anisyl alcohol
Ⓕ coumarin
Ⓖ benzyl salicylate

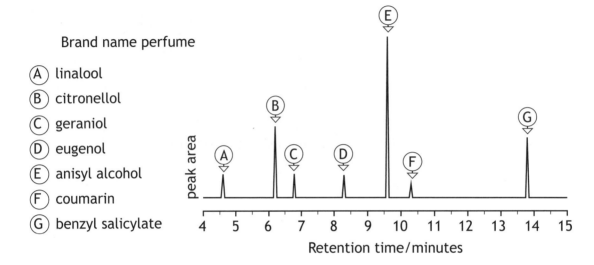

Counterfeit A

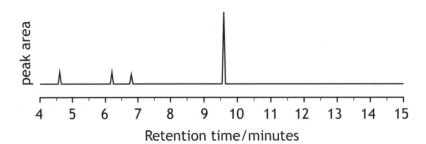

Counterfeit B

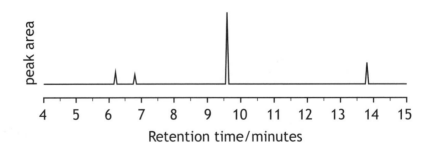

4. (continued)

   (a) Identify one compound present in the brand name perfume that appears in both counterfeit perfumes. **1**

   (b) Some compounds in the brand name perfume are not found in the counterfeit perfumes. State another difference that the chromatograms show between the counterfeit perfumes and the brand name perfume. **1**

   (c) The gas used to carry the perfume sample along the chromatography column is helium.

      (i) Suggest why helium is used. **1**

      (ii) Apart from the polarity of the molecules, state another factor that would affect the retention time of molecules during gas chromatography. **1**

4. (continued)

   (d) Many of the compounds in perfumes are molecules consisting of joined isoprene units.

   (i) State the name that is given to molecules consisting of joined isoprene units. **1**

   (ii) Geraniol is one of the compounds found in perfume. It has the following structural formula and systematic name.

   $$H-\underset{\underset{H}{|}}{\overset{\overset{H}{|}}{C}}-\underset{}{\overset{\overset{CH_3}{|}}{C}}=\underset{\underset{H}{|}}{\overset{\overset{H}{|}}{C}}-\underset{\underset{H}{|}}{\overset{\overset{H}{|}}{C}}-\underset{\underset{CH_3}{|}}{\overset{\overset{H}{|}}{C}}-\underset{}{\overset{\overset{H}{|}}{C}}=\underset{\underset{OH}{|}}{\overset{\overset{H}{|}}{C}}-\underset{\underset{H}{|}}{\overset{\overset{H}{|}}{C}}-H$$

   3,7-dimethylocta-2,6-dien-1-ol

   Linalool can also be present. Its structural formula is shown.

   $$H-\underset{\underset{H}{|}}{\overset{\overset{H}{|}}{C}}-\underset{}{\overset{\overset{CH_3}{|}}{C}}=\underset{\underset{H}{|}}{\overset{\overset{H}{|}}{C}}-\underset{\underset{H}{|}}{\overset{\overset{H}{|}}{C}}-\underset{\underset{H}{|}}{\overset{\overset{OH}{|}}{C}}-\underset{\underset{CH_3}{|}}{\overset{\overset{H}{|}}{C}}=\underset{}{\overset{\overset{H}{|}}{C}}-H$$

   (A) State the systematic name for linalool. **1**

   (B) Explain why linalool can be classified as a tertiary alcohol. **1**

MARKS | DO NOT WRITE IN THIS MARGIN

4. **(continued)**

(e) Coumarin is another compound found in the brand name perfume. It is present in the spice cinnamon and can be harmful if eaten in large quantities.

The European Food Safety Authority gives a tolerable daily intake of coumarin at 0·10 mg per kilogram of body weight.

1·0 kg of cinnamon powder from a particular source contains 4·4 g of coumarin. Calculate the mass of cinnamon powder, in g, which would need to be consumed by an adult weighing 75 kg to reach the tolerable daily intake. **2**

[Turn over

5. Patterns in the Periodic Table

The Periodic Table is an arrangement of all the known elements in order of increasing atomic number. The reason why the elements are arranged as they are in the Periodic Table is to fit them all, with their widely diverse physical and chemical properties, into a logical pattern.

Periodicity is the name given to regularly-occurring similarities in physical and chemical properties of the elements.

Some Groups exhibit striking similarity between their elements, such as Group 1, and in other Groups the elements are less similar to each other, such as Group 4, but each Group has a common set of characteristics.

Adapted from Royal Society of Chemistry, Visual Elements (rsc.org)

Using your knowledge of chemistry, comment on similarities and differences in the patterns of physical and chemical properties of elements in both Group 1 and Group 4.

3

MARKS | DO NOT WRITE IN THIS MARGIN

6. Uncooked egg white is mainly composed of dissolved proteins. During cooking processes, the proteins become denatured as the protein chains unwind, and the egg white solidifies.

(a) Explain why the protein chains unwind. **1**

(b) The temperature at which the protein becomes denatured is called the melting temperature. The melting temperature of a protein can be determined using fluorescence. In this technique, the protein is mixed with a dye that gives out visible light when it attaches to hydrophobic parts of the protein molecule. The hydrophobic parts of the structure are on the inside of the protein and the dye has no access to them unless the protein unwinds.

(i) Ovalbumin is a protein found in egg white. Part of the structure of unwound ovalbumin is shown below.

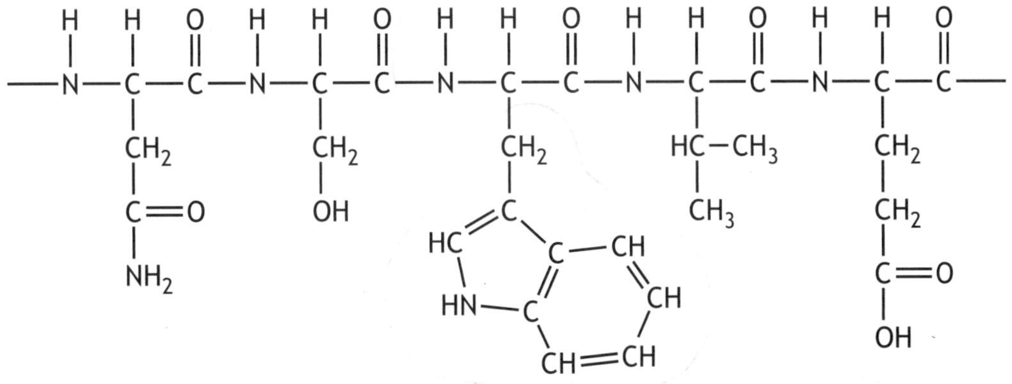

Circle the part of the structure to which the hydrophobic dye is most likely to attach. **1**

[Turn over

6. (b) (continued)

(ii) Another protein in egg white is conalbumin. The temperature of a conalbumin/dye mixture is gradually increased. The fluorescence is measured and a graph is produced.

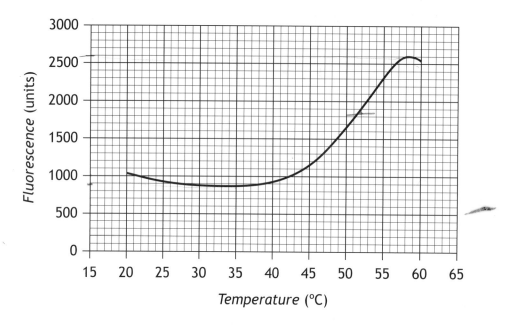

The melting temperature is the temperature at which the fluorescence is halfway between the highest and lowest fluorescence values.

Determine the melting temperature, in °C, for this protein. **1**

MARKS | DO NOT WRITE IN THIS MARGIN

6. (continued)

(c) Once cooked and eaten, the digestive system breaks the protein chains into amino acids with the help of enzymes.

(i) State the name of the digestion process where enzymes break down proteins into amino acids. **1**

(ii)

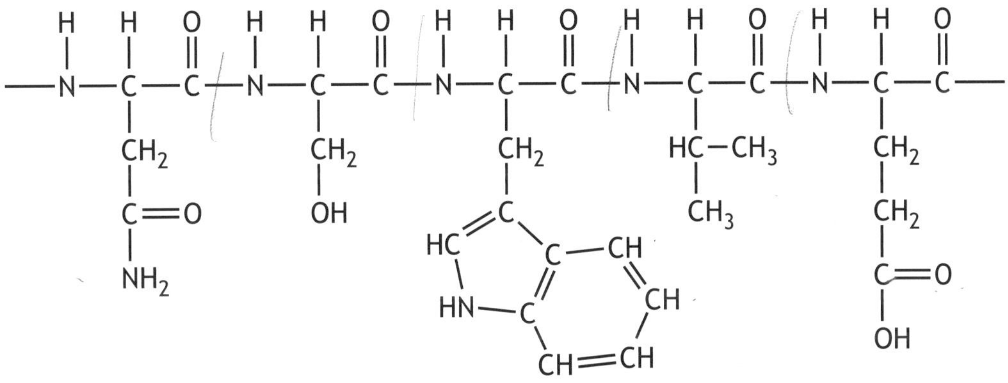

(A) State how many amino acid molecules joined to form this section of protein. **1**

(B) Draw the structure of one amino acid that would be produced when this section of the protein chain is broken down. **1**

7. Methanol can be used as a fuel, in a variety of different ways.

$$\begin{array}{c} H \\ | \\ H-C-OH \\ | \\ H \end{array}$$

(a) An increasingly common use for methanol is as an additive in petrol.

Methanol has been tested as an additive in petrol at 118 g per litre of fuel.

Calculate the volume of carbon dioxide, in litres, that would be released by combustion of 118 g of methanol.

$$2CH_3OH(\ell) + 3O_2(g) \rightarrow 2CO_2(g) + 4H_2O(\ell)$$

(Take the molar volume of carbon dioxide to be 24 litres mol^{-1}). **2**

7. (continued)

(b) A student investigated the properties of methanol and ethanol.

(i) The student carried out experiments to determine the enthalpy of combustion of the alcohols.

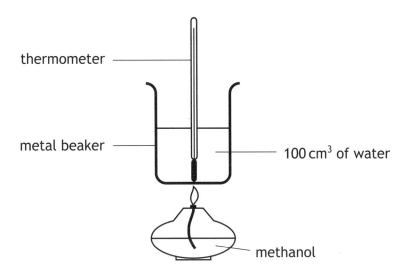

(A) The student carried out the first experiment as shown, but was told to repeat the experiment as the thermometer had been placed in the wrong position.

Suggest why the student's placing of the thermometer was incorrect. **1**

(B) The student always used 100 cm³ of water.

State another variable that the student should have kept constant. **1**

[Turn over

7. (b) (i) (continued)

(C) The student burned 1·07 g of methanol and recorded a temperature rise of 23 °C.

Calculate the enthalpy of combustion, in kJ mol⁻¹, for methanol using the student's results.

3

(ii) The student determined the density of the alcohols by measuring the mass of a volume of each alcohol.

The student's results are shown below.

	Methanol	Ethanol
Volume of alcohol (cm³)	25·0	25·0
Mass of alcohol (g)	19·98	20·05
Density of alcohol (g cm⁻³)		0·802

Calculate the density, in g cm⁻³, of methanol.

1

7. (continued)

(c) Methanol is used as a source of hydrogen for fuel cells. The industrial process involves the reaction of methanol with steam.

$$CH_3OH + H_2O \longrightarrow 3H_2 + CO_2$$

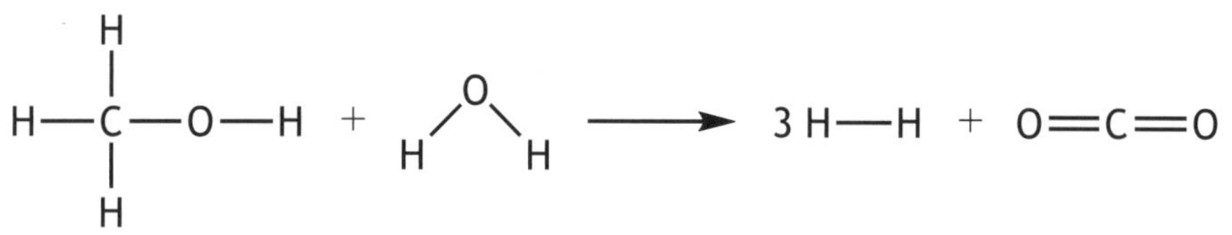

(i) State why it is important for chemists to predict whether reactions in an industrial process are exothermic or endothermic. **1**

(ii) Using bond enthalpies from the data booklet, calculate the enthalpy change, in kJ mol^{-1}, for the reaction of methanol with steam. **2**

[Turn over

8. Sodium carbonate is used in the manufacture of soaps, glass and paper as well as the treatment of water.

One industrial process used to make sodium carbonate is the Solvay process.

The Solvay process involves several different chemical reactions.

It starts with heating calcium carbonate to produce carbon dioxide, which is transferred to a reactor where it reacts with ammonia and brine. The products of the reactor are solid sodium hydrogencarbonate and ammonium chloride which are passed into a separator.

The sodium hydrogencarbonate is heated to decompose it into the product sodium carbonate along with carbon dioxide and water. To recover ammonia the ammonium chloride from the reactor is reacted with calcium oxide produced by heating the calcium carbonate. Calcium chloride is a by-product of the ammonia recovery process.

(a) Using the information above, complete the flow chart by adding the names of the chemicals involved. **2**

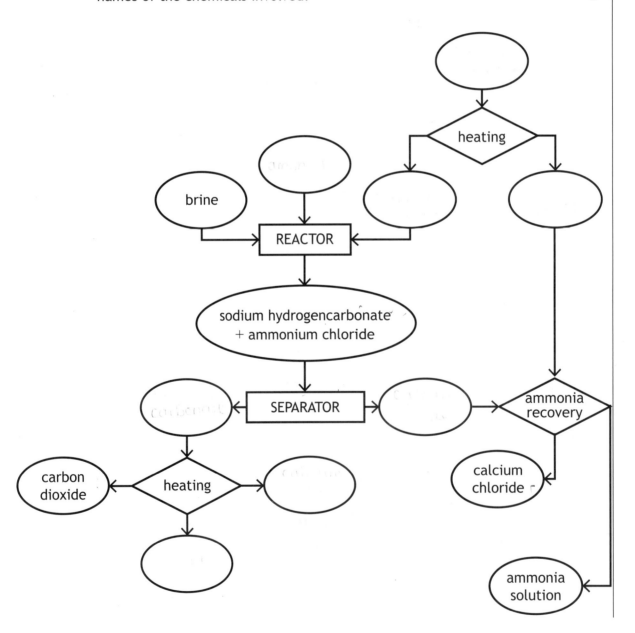

MARKS | DO NOT WRITE IN THIS MARGIN

8 (continued)

(b) The reaction that produces the solid sodium hydrogencarbonate involves the following equilibrium:

$$HCO_3^-(aq) + Na^+(aq) \rightleftharpoons NaHCO_3(s)$$

Brine is a concentrated sodium chloride solution.

Explain fully why using a concentrated sodium chloride solution encourages production of sodium hydrogencarbonate as a solid. **2**

[Turn over

9. Occasionally, seabirds can become contaminated with hydrocarbons from oil spills. This causes problems for birds because their feathers lose their waterproofing, making the birds susceptible to temperature changes and affecting their buoyancy. If the birds attempt to clean themselves to remove the oil, they may swallow the hydrocarbons causing damage to their internal organs.

Contaminated seabirds can be cleaned by rubbing vegetable oil into their feathers and feet before the birds are rinsed with diluted washing-up liquid.

Using your knowledge of chemistry, comment on the problems created for seabirds by oil spills and the actions taken to treat affected birds.

3

[Turn over for Question 10 on *Page twenty-eight*

DO NOT WRITE ON THIS PAGE

10. Plants require trace metal nutrients, such as zinc, for healthy growth. Zinc ions are absorbed from soil through the plant roots.

The zinc ion concentration in a solution can be found by adding a compound which gives a blue colour to the solution with zinc ions. The concentration of zinc ions is determined by measuring the absorption of light by the blue solution. The higher the concentration of zinc ions in a solution, the more light is absorbed.

A student prepared a stock solution with a zinc ion concentration of $1\,g\,l^{-1}$. Samples from this were diluted to produce solutions of known zinc ion concentration.

(a) The stock solution was prepared by adding $1{\cdot}00\,g$ of zinc metal granules to $20\,cm^3$ of $2\,mol\,l^{-1}$ sulfuric acid in a $1000\,cm^3$ standard flask.

$$Zn(s) + H_2SO_4(aq) \rightarrow ZnSO_4(aq) + H_2(g)$$

The flask was left for 24 hours, without a stopper. The solution was then diluted to $1000\,cm^3$ with water.

(i) **Explain fully** why the flask was left for 24 hours, without a stopper. **2**

(ii) Explain why the student should use deionised water or distilled water, rather than tap water, when preparing the stock solution. **1**

(b) Solutions of known zinc ion concentration were prepared by transferring accurate volumes of the stock solution to standard flasks and diluting with water.

(i) Name the piece of apparatus which should be used to transfer $10\,cm^3$ of stock solution to a standard flask. **1**

10. (b) (continued)

(ii) Calculate the concentration, in mg l^{-1}, of the solution prepared by transferring 10 cm^3 of the 1 g l^{-1} stock solution to a 1000 cm^3 standard flask and making up to the mark.

1

(c) The light absorbance of different solutions was measured and the results plotted.

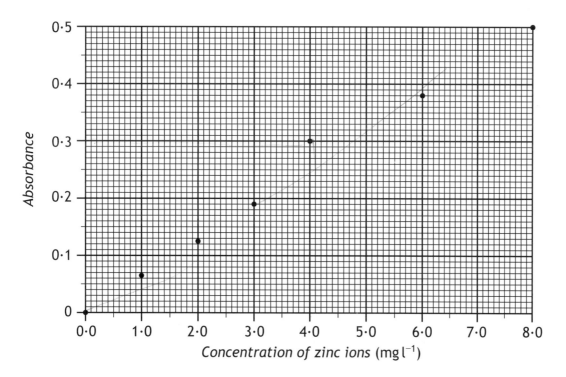

A solution prepared from a soil sample was tested to determine the concentration of zinc ions. The solution had an absorbance of 0·3.

Determine the concentration, in mg l^{-1}, of zinc ions in the solution.

1

MARKS | DO NOT WRITE IN THIS MARGIN

11.

4-hydroxybenzoic acid

4-hydroxybenzoic acid can react with alcohols to form compounds known as parabens.

(a) Name the functional group circled in the structure of 4–hydroxybenzoic acid. **1**

(b) Name the type of reaction taking place when parabens are formed. **1**

(c) Draw the paraben formed when 4–hydroxybenzoic acid reacts with ethanol. **1**

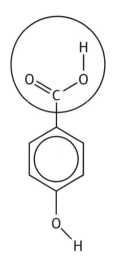

MARKS DO NOT WRITE IN THIS MARGIN

11. (continued)

(d) Parabens can be used as preservatives in cosmetics and toiletries.

Parabens are absorbed into the body through the skin. The following table indicates the absorption of some parabens.

Paraben	*Absorption* ($\mu g\,cm^{-2}$)
Methyl	32·50
Ethyl	20·74
Propyl	11·40
Butyl	7·74
Hexyl	1·60

State a conclusion that can be drawn from the information in the table. **1**

[Turn over

12. (a) The concentration of sodium hypochlorite in swimming pool water can be determined by redox titration.

Step 1

A 100·0 cm^3 sample from the swimming pool is first reacted with an excess of acidified potassium iodide solution forming iodine.

NaOCl(aq) + 2I$^-$(aq) + 2H$^+$(aq) → I$_2$(aq) + NaCl(aq) + H$_2$O(ℓ)

Step 2

The iodine formed in step 1 is titrated using a standard solution of sodium thiosulfate, concentration 0·00100 mol l^{-1}. A small volume of starch solution is added towards the endpoint.

I$_2$(aq) + 2Na$_2$S$_2$O$_3$(aq) → 2NaI(aq) + Na$_2$S$_4$O$_6$(aq)

(i) Describe in detail how a burette should be prepared and set up, ready to begin the titration. **3**

(ii) Write the ion-electron equation for the oxidation reaction occurring in step 1. **1**

MARKS | DO NOT WRITE IN THIS MARGIN

12. (a) (continued)

(iii) Calculate the concentration, in $mol\,l^{-1}$, of sodium hypochlorite in the swimming pool water, if an average volume of $12{\cdot}4\,cm^3$ of sodium thiosulfate was required. **3**

(b) The level of hypochlorite in swimming pools needs to be maintained between 1 and 3 parts per million (1 – 3 ppm).

400 cm^3 of a commercial hypochlorite solution will raise the hypochlorite level of 45 000 litres of water by 1 ppm.

Calculate the volume of hypochlorite solution that will need to be added to an Olympic-sized swimming pool, capacity 2 500 000 litres, to raise the hypochlorite level from 1 ppm to 3 ppm. **2**

[Turn over

12. (continued)

(c) The familiar chlorine smell of a swimming pool is not due to chlorine but compounds called chloramines. Chloramines are produced when the hypochlorite ion reacts with compounds such as ammonia, produced by the human body.

$$OCl^-(aq) + NH_3(aq) \rightarrow NH_2Cl(aq) + OH^-(aq)$$
monochloramine

$$OCl^-(aq) + NH_2Cl(aq) \rightarrow NHCl_2(aq) + OH^-(aq)$$
dichloramine

$$OCl^-(aq) + NHCl_2(aq) \rightarrow NCl_3(aq) + OH^-(aq)$$
trichloramine

Chloramines are less soluble in water than ammonia due to the polarities of the molecules, and so readily escape into the atmosphere, causing irritation to the eyes.

(i) Explain the difference in polarities of ammonia and trichloramine molecules.

ammonia   trichloramine

2

MARKS DO NOT WRITE IN THIS MARGIN

12. (c) (continued)

(ii) Chloramines can be removed from water using ultraviolet light treatment.

One step in the process is the formation of free radicals.

$$NH_2Cl \xrightarrow{UV} \bullet NH_2 + \bullet Cl$$

State what is meant by the term free radical. **1**

(iii) Another step in the process is shown below.

$$NH_2Cl + \bullet Cl \longrightarrow \bullet NHCl + HCl$$

State the name for this type of step in a free radical reaction. **1**

[Turn over for Question 13 on *Page thirty-six*

13. (a) One test for glucose involves Fehling's solution.

Circle the part of the glucose molecule that reacts with Fehling's solution.

$$H-\underset{\underset{H}{|}}{\overset{\overset{OH}{|}}{C}}-\underset{\underset{OH}{|}}{\overset{\overset{H}{|}}{C}}-\underset{\underset{OH}{|}}{\overset{\overset{H}{|}}{C}}-\underset{\underset{H}{|}}{\overset{\overset{OH}{|}}{C}}-\underset{\underset{OH}{|}}{\overset{\overset{H}{|}}{C}}-\overset{\overset{H}{|}}{\underset{\underset{O}{\|}}{C}}-H$$

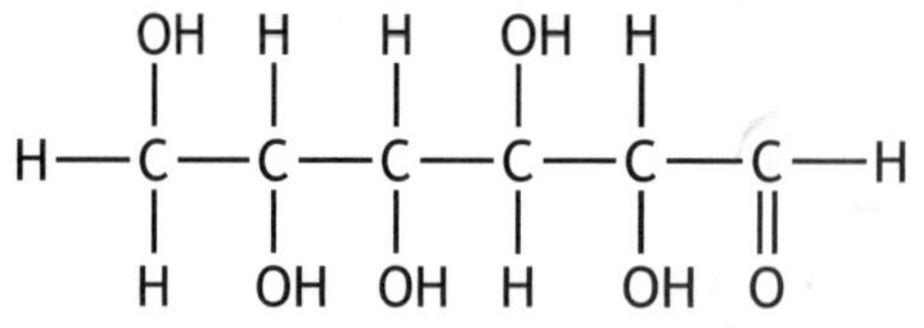

(b) In solution, sugar molecules exist in an equilibrium in straight-chain and ring forms.

To change from the straight-chain form to the ring form, the oxygen of the hydroxyl on carbon number 5 joins to the carbonyl carbon. This is shown below for glucose.

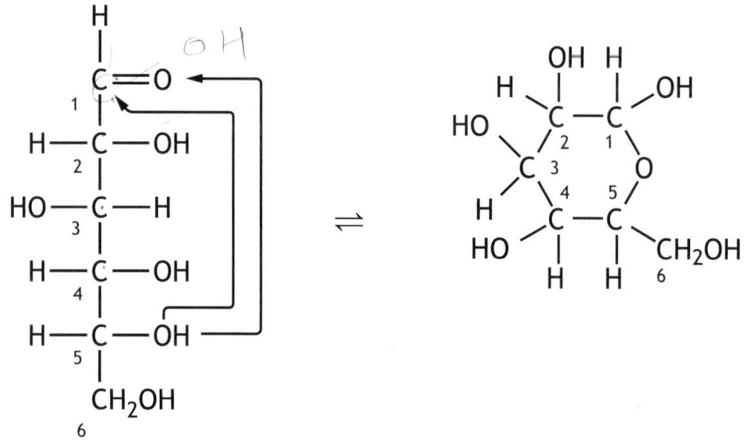

glucose

Draw the structure of a ring form for fructose.

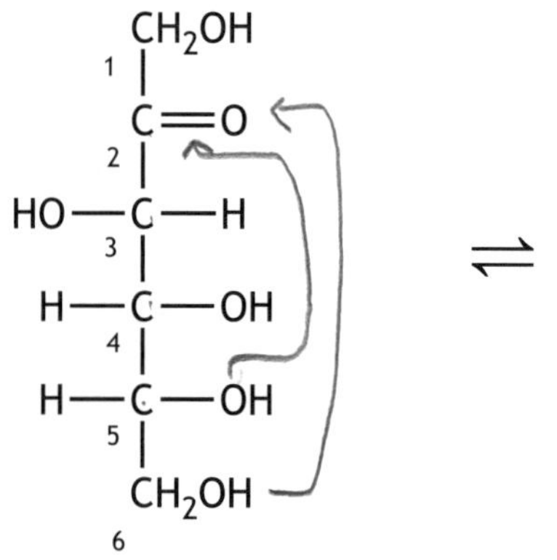

fructose

[END OF QUESTION PAPER]

ADDITIONAL DIAGRAM FOR USE IN QUESTION 3 (a) (i)

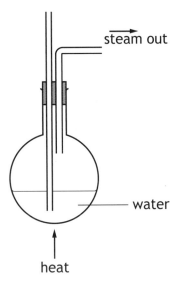

ADDITIONAL SPACE FOR ANSWERS AND ROUGH WORK

ADDITIONAL SPACE FOR ANSWERS AND ROUGH WORK

HIGHER FOR CfE | ANSWER SECTION

ANSWER SECTION FOR

SQA AND HODDER GIBSON HIGHER FOR CfE CHEMISTRY 2015

HIGHER FOR CfE CHEMISTRY SPECIMEN QUESTION PAPER

Section 1

Question	Response
1.	C
2.	C
3.	A
4.	D
5.	A
6.	A
7.	D
8.	D
9.	B
10.	C
11.	C
12.	A
13.	C
14.	B
15.	D
16.	B
17.	B
18.	A
19.	B
20.	D

Section 2

1. (a) (i) Na(g) → Na$^+$(g) + e$^-$
   No necessity to show negative charge on e
   Maximum mark: 1

   (ii) Idea that ionisation energy is removal of an electron (1 mark)

   Idea that 1st ionisation energy is removal of electron from 3rd (outermost) shell and second is removal of electron from an inner shell. (1 mark)

   Idea of shielding effect of inner electrons ie that second electron is less well shielded from nuclear pull and therefore more energy is needed to remove electron.

   OR

   The removal from a full shell requires more energy than removal from an incomplete shell. (1 mark)

   Maximum mark: 3

   (b) 0·178 g (0·18 g)   **Maximum mark: 1**

2. (a) Molecules must collide with energy greater than activation energy (sufficient energy to react) (1 mark)

   and molecules must collide with correct orientation (1 mark)

   Energy mark must convey the idea that there is a minimum energy required for the molecules to react
   Maximum mark: 2

   (b) (i) 2 moles acid give 2 moles gas (1 mark)
   Answer = 0·00041 (0·0004) mol (1 mark)
   Maximum mark: 2

   (ii)

Reaction step	Name of step
Cl$_2$ → 2Cl•	Initiation (1 mark)
Cl• + H$_2$ → HCl + H•	propagation
H• + Cl• → HCl (1 mark) (or any appropriate reaction step in which two free radicals combine to give a molecule)	termination

   Maximum mark: 2

   (iii) Correct bond enthalpies selected (436; 243; 432 kJ mol^{-1}) (1 mark)
   Answer = −185

   Negative sign required in the answer
   Maximum mark: 2

3. (a) (i) Butyl propanoate (1 mark)   **Maximum mark: 1**

   (ii) B > A > C (1 mark)   **Maximum mark: 1**

   (b)
   ```
        H          (CH₂)₇
         \        /
          C
          ||           HC — OH
          C
         /        \
        H          (CH₂)₇
   ```

   H must be shown on C to which hydroxyl group is attached.   **Maximum mark: 1**

   (c)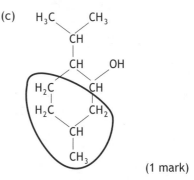

   (1 mark)

   Any group of 5 carbons with 1 branch. OH to be ignored.   **Maximum mark: 1**

4. There may be strengths and weaknesses in the candidate response: *assessors should focus as far as possible on the strengths, taking account of weaknesses (errors or omissions) only where they detract from the overall answer in a significant way, which should then be taken into account when determining whether the response demonstrates **reasonable**, **limited** or **no** understanding.*

This open-ended question requires comment on **the changes to food that may occur during cooking.** Candidate responses are expected to make comment on the basis of relevant chemistry ideas/concepts which might include one or more of: denaturing proteins; solubility of flavouring molecules; effect on antioxidants; emulsification; or other relevant ideas/concepts.

Note that 'cooking' should be taken to mean any steps in a recipe.

3 marks: The candidate has demonstrated a good conceptual understanding of the chemistry involved, providing a logically correct response to the problem/situation presented. This type of response might include a statement of principle(s) involved, a relationship or equation, and the application of these to respond to the problem/situation. This does not mean the answer has to be what might be termed an 'excellent' answer or a 'complete' one.	In response to this question, a **good** understanding might be demonstrated by a candidate response that: • makes comments based on one relevant chemistry idea/concept, in a **detailed/developed** response that is **correct or largely correct** (any weaknesses are minor and do not detract from the overall response), **OR** • makes comments based on a range of relevant chemistry ideas/concepts, in a response that is **correct or largely** correct (any weaknesses are minor and do not detract from the overall response), **OR** • otherwise demonstrates a good understanding of the chemistry involved.
2 marks: The candidate has demonstrated a **reasonable** understanding of the chemistry involved, showing that the problem/situation is understood. This type of response might make some statement(s) that is/are relevant to the problem/situation, for example, a statement of relevant principle(s) or identification of a relevant relationship or equation.	In response to this question, a **reasonable** understanding might be demonstrated by a candidate response that: • makes comments based on one or more relevant chemistry idea(s)/concept(s), in a response that is **largely correct** but has **weaknesses** which detract to a small extent from the overall response, **OR** • otherwise demonstrates a reasonable understanding of the chemistry involved.
1 mark: The candidate has demonstrated a limited understanding of the chemistry involved, showing that a little of the chemistry that is relevant to the problem/situation is understood. The candidate has made some statement(s) that is/are relevant to the problem/situation.	In response to this question, a **limited** understanding might be demonstrated by a candidate response that: • makes comments based on one or more relevant chemistry idea(s)/concept(s), in a response that has **weaknesses** which detract to a large extent from the overall response, **OR** • otherwise demonstrates a limited understanding of the chemistry involved.
0 marks: The candidate has demonstrated no understanding of the chemistry that is relevant to the problem/situation. The candidate has made no statement(s) that is/are relevant to the problem/situation.	Where the candidate has only demonstrated knowledge and understanding of chemistry **that is not relevant to the problem/situation presented**, 0 marks should be awarded.

5. (a) (351 000/36·1) * 1·25 (1 mark)
= 12 154 cm³ or 12·15 litres
1 mark for correctly calculated figure
1 mark for correct units **Maximum mark: 3**

(b) [Structural diagram of propan-1-ol with hydrogen bonding to water molecules]

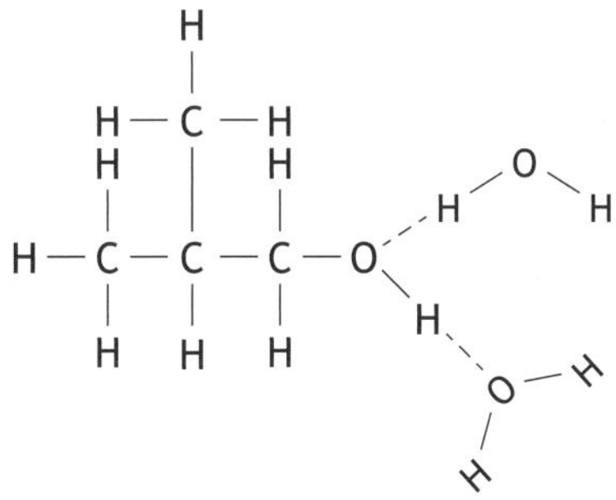

One or the other hydrogen bonds may be shown

Only one water molecule with the hydrogen bond need be shown. **Maximum mark: 1**

(c) ΔH = [+ 335 + (–17) + (–242)] (1 mark)
= (+) 76 (kJ mol⁻1) (1 mark)

Sign and units need not be given in answer but if given must be correct. **Maximum mark: 2**

(d) 1 mark for design of experiment. Eg inverting sealed tubes and measuring time taken for an air bubble to rise through the tube or dropping a ballbearing into the fuels and timing how long they take to drop or any other reasonable experiment.

1 mark for stating how experiment allows viscosities to be compared. **Maximum mark: 2**

6. (a) [Structural diagram of ethyl-2-cyanoacrylate with ester group circled]

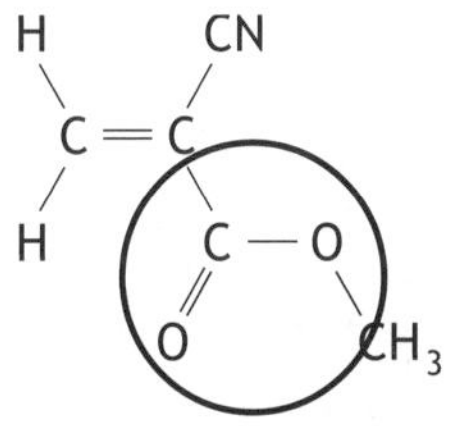

(1 mark) **Maximum mark: 1**

(b) (i) moisture/water on the surface of the skin/ cyanoacrylate monomers rapidly polymerise in the presence of water (1 mark) **Maximum mark: 1**

(ii) permanent dipole – permanent dipole attractions (1 mark) **Maximum mark: 1**

(c) (i) Methanal (1 mark) **Maximum mark: 1**

(ii) Condensation (1 mark) **Maximum mark: 1**

(iii) Atom economy = [mass of ethyl-2-cyanoacrylate/total mass of reactants] × 100 (1 mark)
= [125 / 143] x100
= 87·41% (1 mark) **Maximum mark: 2**

(d) Value between 420 and 1140 Ncm⁻² (1 mark)

units not required **Maximum mark: 1**

(e) (i) [Structural diagram of cyanoacrylate with C_8H_{17} side chain]

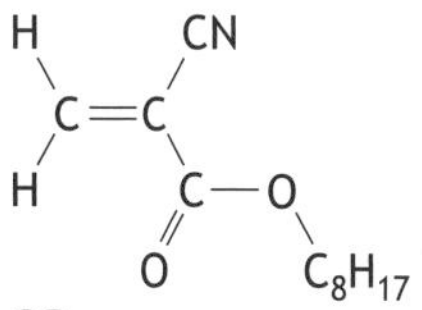

OR

[Structural diagram with $CH_2CH_2CH_2CH_2CH_2CH_2CH_2CH_3$ side chain]

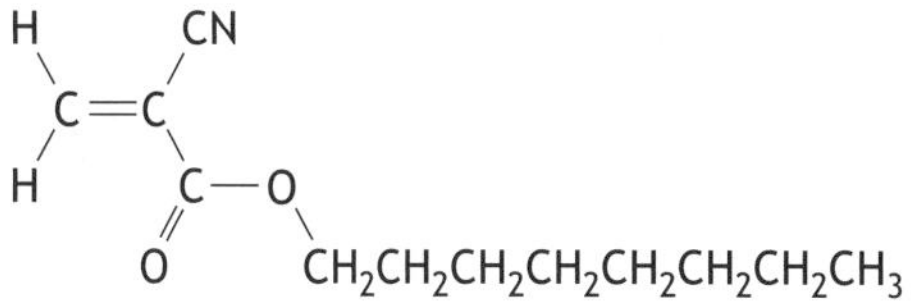

(1 mark)

accept any saturated C-8 side-chain **Maximum mark: 1**

(ii) Less heat is given out/less exothermic reaction using the octyl/butyl mix so less damage/pain/ burning will occur to the wound/patient's skin

OR

the reaction takes place more quickly sealing the wound (1 mark) **Maximum mark: 1**

7. (a) (i) A solution of accurately known concentration (1 mark) **Maximum mark: 1**

(ii) The weighed sample is dissolved in a small volume of (deionised) water in a beaker and the solution transferred to a standard flask. The beaker is rinsed and the rinsings also poured into the standard flask. (1 mark)

The flask is made up to the mark adding the last few drops of water using a dropping pipette. The flask is stoppered and inverted several times to ensure thorough mixing of the solution.(1 mark) **Maximum mark: 2**

(iii) Tap water contains dissolved salts (1mark) that may react with sodium fluoride and affect the concentration of the solution. (1 mark) **Maximum mark: 2**

(iv) 2 mg l⁻¹ (1 mark) **Maximum mark: 1**

(b) (i) Moles MnO₄⁻ (aq) = 0·00034 mol (1 mark)
Moles NO₂⁻ (aq) in 25 cm³ = 0·0081 mol (1 mark)
Concentration = 0·0324 mol l⁻¹ (1 mark)
Maximum mark: 3

(ii) NO₂⁻ (aq) + H₂O(l) → NO₃⁻ (aq) + 2H⁺(aq) + 2e⁻ (1 mark)

State symbols and charge on electron are not essential **Maximum mark: 1**

8. (a) Carboxyl (group) (1 mark) **Maximum mark: 1**

(b) (i) Large hydrocarbon section attached to the carboxyl group making this section insoluble in water. (1 mark) **Maximum mark: 1**

(ii) [Structural diagram with fatty acid chain circled: $CH_2-O-C-(CH_2)_6-CH=CH-(CH_2)_7-CH_3$]

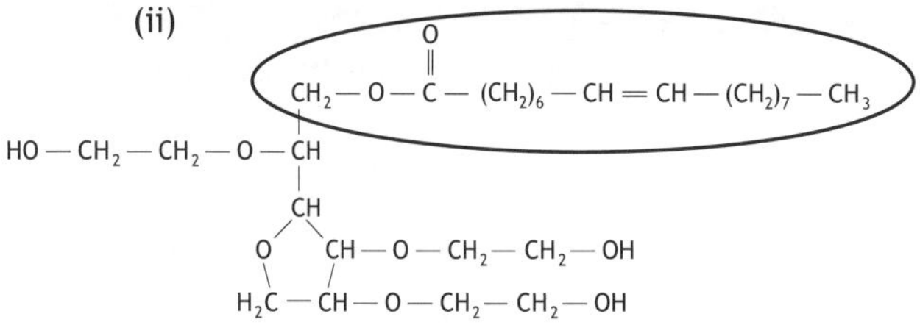

(1 mark)

Identified portion does not need to include the ester link. **Maximum mark: 1**

(iii) 2·5 cm³ (1 mark) **Maximum mark: 1**

(c) (i) Amide/peptide (link) (1 mark) **Maximum mark: 1**

(ii) 0·023(0·0225) **Maximum mark: 1**

9. (a) (i) Shea butter has fewer double bonds/is not very unsaturated (1 mark)

The London dispersion forces or van der Waals' forces between its molecules are stronger than in oils, therefore melting point higher. (1 mark)
Maximum mark: 2

(ii) 85-171

Actual answer is approx. 134 **Maximum mark: 1**

(iii) oxygen **Maximum mark: 1**

(b) (i) Glycerol/propane-1,2,3-triol **Maximum mark: 1**

(ii) Waterbath/Heating mantle
Bunsen burner is not acceptable
Maximum mark: 1

(iii) = 24·8/25% (2 marks)
Calculates the theoretical yield of soap (= 5·16g)

OR

correctly calculates the number of moles of reactant (= 0·00566) and product (= 0·00421) [1·28/304] (1 mark)

Calculating the % yield; either using the actual and theoretical masses, or using the actual number of moles of products and actual number of moles of reactant (1 mark)

Not acceptable for % yield mark to use 1·28/5 or 3 × 1·28/5. **Maximum mark: 3**

10. (a) (i) 2 mol NH_3 gives 1 mol $(NH_2)_2CO$ (1 mark)

OR

34 g gives 60 g (1 mark)

300 tonnes (1 mark)

Units not required **Maximum mark: 2**

(ii) An excess of ammonia will push the equilibrium to the left (1 mark) **Maximum mark: 1**

(iii) Low pressure favours the reaction that produces gas molecules thereby increasing the pressure.

OR

there are more moles of gas on the right-hand side of the equation (1 mark)

The forward reaction will be favoured by low pressure causing the carbamate to break down (1 mark) **Maximum mark: 2**

(iv) A line from the ammonia carbon dioxide coming from the separator back to the Reactors (1 mark)
Maximum mark: 1

(b) (i) −88 kJ mol^{-1} (1 mark)
Negative sign required; units not required

(ii)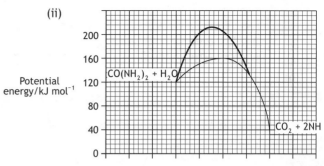
(1 mark)
Curve showing higher activation energy
Maximum mark: 1

11. There may be strengths and weaknesses in the candidate response: *assessors should focus as far as possible on the strengths, taking account of weaknesses (errors or omissions) only where they detract from the overall answer in a significant way, which should then be taken into account when determining whether the response demonstrates **reasonable**, **limited** or **no** understanding.*

This open-ended question requires comment on **what compounds the old pharmacy jars might contain now**. Candidate responses are expected to make comment, on the basis of relevant chemistry ideas/concepts which might include one or more of: reactions of molecules with ethanol; or reaction of molecules with each other; or with oxygen if this has entered; or length of time for reactions to happen; or other relevant ideas/concepts.

Note that 'cooking' should be taken to mean any steps in a recipe.

| 3 marks: The candidate has demonstrated a good conceptual understanding of the chemistry involved, providing a logically correct response to the problem/situation presented. This type of response might include a statement of principle(s) involved, a relationship or equation, and the application of these to respond to the problem/situation. This does not mean the answer has to be what might be termed an 'excellent' answer or a 'complete' one. | In response to this question, a **good** understanding might be demonstrated by a candidate response that:
• makes comments based on one relevant chemistry idea/concept, in a **detailed/developed** response that is **correct or largely correct** (any weaknesses are minor and do not detract from the overall response), **OR**
• makes comments based on a range of relevant chemistry ideas/concepts, in a response that is **correct or largely** correct (any weaknesses are minor and do not detract from the overall response), **OR**
• otherwise demonstrates a good understanding of the chemistry involved. |

2 marks: The candidate has demonstrated a **reasonable** understanding of the chemistry involved, showing that the problem/situation is understood. This type of response might make some statement(s) that is/are relevant to the problem/situation, for example, a statement of relevant principle(s) or identification of a relevant relationship or equation.	In response to this question, a **reasonable** understanding might be demonstrated by a candidate response that: • makes comments based on one or more relevant chemistry idea(s)/concept(s), in a response that is **largely correct** but has **weaknesses** which detract to a small extent from the overall response, **OR** • otherwise demonstrates a reasonable understanding of the chemistry involved.
1 mark: The candidate has demonstrated a limited understanding of the chemistry involved, showing that a little of the chemistry that is relevant to the problem/situation is understood. The candidate has made some statement(s) that is/are relevant to the problem/situation.	In response to this question, a **limited** understanding might be demonstrated by a candidate response that: • makes comments based on one or more relevant chemistry idea(s)/concept(s), in a response that has **weaknesses** which detract to a large extent from the overall response, **OR** • otherwise demonstrates a limited understanding of the chemistry involved.
0 marks: The candidate has demonstrated no understanding of the chemistry that is relevant to the problem/situation. The candidate has made no statement(s) that is/are relevant to the problem/situation.	Where the candidate has only demonstrated knowledge and understanding of chemistry **that is not relevant to the problem/situation presented**, 0 marks should be awarded.

12. (a) An amino acid that must be obtained through our diet

    OR

    cannot be synthesised by the body (1 mark)

    Maximum mark: 1

    (b) RDA (60 kg adult) = 900 mg (1 mark)
    Mass of tuna = 119 (119·2) g (1 mark)

    Maximum mark: 1

(c) (i) Tyrosine (1 mark) **Maximum mark: 1**

(ii)

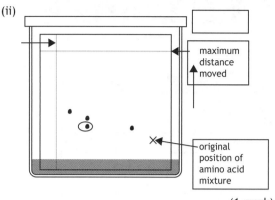

(1 mark)

Maximum mark: 1

(iii) With solvent 1, alanine and threonine have the same Rf value and travel the same distance and show as a single spot. When they are placed in solvent 2 the spot splits into 2 since alanine and threonine have different Rf values. (1 mark)

Maximum mark: 1

HIGHER FOR CfE CHEMISTRY MODEL PAPER 1

Section 1

Question	Response
1.	B
2.	C
3.	C
4.	C
5.	B
6.	D
7.	A
8.	B
9.	B
10.	C
11.	D
12.	B
13.	B
14.	D
15.	A
16.	D
17.	C
18.	D
19.	A
20.	C

Section 2

1. (a) (i) Any metal element

 (ii) Covalent network

 (iii) Phosphorus or Sulfur or Carbon (fullerene)

 (iv) Monatomic

 Maximum mark: All correct – 2
 2 or 3 correct – 1

 (b) (i) Li(g) ⟶ Li$^+$(g) + e$^-$

 State symbols are required. This equation comes from the data booklet.

 Maximum mark: 1

 (ii) Idea that ionisation energy is removal of an electron (1 mark)

 Idea that 1st ionisation energy is removal of electron from 2nd (outermost) shell and second is removal of electron from an inner shell. (1 mark)

 Idea of shielding effect of inner electrons ie that second electron is less well shielded from nuclear pull and therefore more energy is needed to remove electron.

 OR

 The removal from a full shell requires more energy than removal from an incomplete shell. (1 mark)

 Ionisation energy is affected by the nuclear charge and the screening from electron shells.

 Maximum mark: 3

 (iii) 75 (mg)
 2 × 6.9 = 13.8
 $\frac{13.8}{73.8}$ × 400 = 75

 Maximum mark: 1

 (c) This is an open-ended question

 The whole candidate response should first be read to establish its overall quality in terms of accuracy and relevance to the problem/situation presented. There may be strengths and weaknesses in the candidate response: assessors should focus as far as possible on the strengths, taking account of weaknesses (errors or omissions) only where they detract from the overall answer in a significant way, which should then be taken into account when determining whether the response demonstrates reasonable, limited or no understanding. Assessors should use their professional judgement to apply the guidance below to the wide range of possible candidate responses.

 3 marks: The candidate has demonstrated a good conceptual understanding of the chemistry involved, providing a logically correct response to the problem/situation presented.

 This type of response might include a statement of principle(s) involved, a relationship or equation, and the application of these to respond to the problem/situation. This does not mean the answer has to be what might be termed an "excellent" answer or a "complete" one.

 2 marks: The candidate has demonstrated a reasonable understanding of the chemistry involved, showing that the problem/situation is understood.

 This type of response might make some statement(s) that is/are relevant to the problem/situation, for example, a statement of relevant principle(s) or identification of a relevant relationship or equation

 1 mark: The candidate has demonstrated a limited understanding of the chemistry involved, showing that a little of the chemistry that is relevant to the problem/situation is understood.

 The candidate has made some statement(s) that is/are relevant to the problem/situation.

 0 marks: The candidate has demonstrated no understanding of the chemistry that is relevant to the problem/situation.

 The candidate has made no statement(s) that is/are relevant to the problem/situation.

 Points you could discuss:

 The reactivity of hydrogen.

 The number of electrons.

 The molecular structure of hydrogen and its low mp/bp.

 Compare to the halogens.

 Maximum mark: 3

2. (a) Stating that one (NH$_3$) is polar and/or the other (NCl$_3$) is non-polar (1 mark)

 Identifying that NH$_3$ has hydrogen bonding and identifying that NCl$_3$ has London dispersion forces (1 mark)

 Other mark is for a statement linking intermolecular forces/polarity to the solubility in water.

Statements such as the following would be acceptable
- Water is polar
- Water can hydrogen bond
- Water is a good solvent for polar molecules
- Like dissolves like

If you are asked about physical properties such as boiling/melting points or solubility, look at the intermolecular bonding of the molecules involved.

Maximum mark: 3

(b) NCl_3 (g) + $3H_2O$ (l) → NH_3 (aq) + 3HOCl (aq)

Maximum mark: 1

(c) (i) Atoms/molecules with unpaired electrons.

Maximum mark: 1

(ii) initiation

Maximum mark: 1

(iii) Line from reactant to products clearly demonstrating a lower Ea.

Maximum mark: 1

(d) $-75 kJ mol^{-1}$

1 mark for any of the following:
945, 3 × 436, 3 × 388, 6 × 388, 2328, 2253

Bonds broken =
(NN) = 945
3(HH) = 3 × 346 = 1308
Total = +2253.

Bonds made = 2 × (3 × NH) = 6 × 388 = −2328
2253 − 2328 = −75

Maximum mark: 2

(e) Correct collision geometry
OR
Must have energy greater than or equal to Ea

Maximum mark: 1

3. (a) (i) The enzyme changes shape when heated

Maximum mark: 1

(ii) Oxygen to hydrogen ratio has decreased
OR
Hydrogen has been gained.

Maximum mark: 1

(iii) 87%

Partial marks can be awarded using a scheme of two "concept" marks and one "arithmetic" mark.

1 mark is given for candidate showing understanding of the concept of an actual yield divided by a theoretical yield either using masses or moles of reactant and product.

1 mark is given for candidate displaying understanding of the 1:2 stoichiometry in the reaction.

1 mark is awarded for correct arithmetic throughout the calculation. This mark can only be awarded if both concept marks have been awarded.

180g ➡ 2 × 46g
46180 = 0.511
1000g ➡ 511g (theoretical)
× 100
$\frac{445}{511}$ × 100 = 87%

Maximum mark: 3

(b) 29717 or roundings

(no units required and ignore sign if included)

$-1367 kJ mol^{-1}$
1 mol of ethanol = 46g i.e.
46g ➡ −1367 kJ
1g ➡ 29.72 kJ
1000g ➡ $29717 kJ kg^{-1}$

Maximum mark: 1

(c) 3·87 (%)

For 1 mark candidate must have either 1035−1005 or 30 or 0·129

Change in SG = 1035−1005 = 30
So, f = 0.129
30 × 0.129 = 3.87

Maximum mark: 2

4. (a)
$$H_2C=C(CH_3)-C(=CH_2)-H$$
(structure of 2-methyl-1,3-butadiene / isoprene)

Maximum mark: 1

(b) Geraniol has hydrogen bonding
OR
there are stronger intermolecular bonds (in geraniol)
OR
stronger van der Waals' (in geraniol)
OR
limonene only has London dispersion forces

Always look at the intermolecular bonding. Here, geraniol has an −OH group. This gives H bonding. Limonene is a hydrocarbon which will have LDF between molecules.

Maximum mark: 1

(c) (i) aldehydes or alkanals

Maximum mark: 1

(ii) $H_3C-(CH_2)_8-CH(CH_3)-C(=O)-OH$

Maximum mark: 1

5. Open-ended question

Points you could discuss:

Emulsifier structures and their function. Why they are necessary in ice-cream (discuss solubility of fats)

The structure of the flavouring ester and its properties.

Purpose of an antioxidant and how this relates to foods spoiling.

Maximum mark: 3

6. (a) (i) Amide or peptide

Maximum mark: 1

(ii) $HO-C(=O)-CH(NH_2)-CH_2-C(=O)-OH$

Maximum mark: 1

(iii) Essential   1

Maximum mark: 1

(b) (i) 69 – 70 (mg l⁻¹) no units required. Ignore incorrect units)

Maximum mark: 1

(ii) Sample of Y should be diluted or less of sample Y should be used or smaller sample of Y

Maximum mark: 1

7. (a) Diagram completed to show viable method of drying gas using calcium oxide.

Maximum mark: 1

(b) (i) 37·7 g
(no units required — ignore incorrect units)

Partial marks can be awarded using a scheme of two "concept" marks and one "arithmetic" mark.

1 mark for demonstration of use of the relationship Eh = cmΔT this mark is for the concept, do not penalise for incorrect units or incorrect arithmetic.

The value of 43·89 (kJ) would automatically gain this mark.

1 mark for demonstration of knowledge that the enthalpy value provided relates to 1 mole of calcium oxide reacting with water.

This mark could be awarded if the candidate is seen to be working out the number of moles of calcium oxide required (0·67) or if the candidate's working shows a proportion calculation involving use of the gfm for calcium oxide (56).

1 mark is awarded for correct arithmetic throughout the calculation. This mark can only be awarded if both concept marks have been awarded.

Eh = cmΔT
= 4.18 × 0.21 × 50 = 43.89 kJ
65 kJ ➡ 1 mol CaO
65kJ ➡ 56g
1kJ ➡ 0.861 g
43.89kJ ➡ 37.81g

Maximum mark: 3

(ii) – 147 kJ mol⁻¹

partial marks

1 mark is awarded for 2 out of the four following numbers being shown
+ 635 + 286 – 986 – 82

Maximum mark: 2

8. (a) (i) Water bath/heating mantle

Maximum mark: 1

(ii) Condensation

Maximum mark: 1

(iii) [structural formula of benzoate ester: phenyl-C(=O)-O-CH₂-CH₃]

Maximum mark: 1

(b) 82·3 (82%)

1 mark: Concept atom economy ie desired product mass over reactant masses (to be exemplified at central marking)

1 mark: Correct arithmetic

$$\frac{2 \times 144}{(2 \times 122) + 106} \times 100 = 82.3$$

Maximum mark: 1

(c) £93.75

Partial marks:
583g **AND** 467g (1 mark)
£92.16 **OR** £1.59 (1 mark)

Benzoic acid:
300g requires 350g

$\frac{350}{300} = 1.167$

500g requires 1.167 × 500 = 583.33g of benzoic acid.
100g costs £15.80

$\frac{15.80}{100} = £0.158$

583.33g costs £92.17

Sodium carbonate:
300g requires 280g

$\frac{280}{300} = 0.933$

500g requires 0.933 × 500 = 466.67 g
1000g costs £3.40

$\frac{3.40}{1000} = £0.0034$

466.67g costs £1.59

Maximum mark: 2

9. (a) Secondary (or tertiary) alcohols have lower boiling points than primary or words to that effect (1 mark)

more branched the (isomeric) alcohol the lower the boiling point (1 mark)

Maximum mark: 2

(b) 121 < b.pt < 149

Maximum mark: 1

10. (a)
```
     O
     ‖
     C—OH
     |
  H—C—OH
     |
  H—C—OH
     |
     C—OH
     ‖
     O
```

Maximum mark: 1

(b) 0.0165 g

1 mark for stoichiometric understanding of NaHCO₃ and CO₂

1 mark for showing conversion to of grams and litres.

1 mark is awarded for correct arithmetic calculation of mass equivalent to 105cm³

$\frac{105}{24000} = 0.004375$

$\frac{0.004375}{2} = 0.0022$

Mass tartaric = 0.0022 × 150 = 0.33

$\frac{0.33}{20} = 0.0164g$

Maximum mark: 3

11. (a) any suitable indication of point at which curves start to level off on concentration axis, eg by a vertical line or arrow

   Equilibrium is reached when the rate of the forward reaction = rate of the reverse reaction. Because of this, the concentration of reactants and products stays constant at equilibrium.

   Maximum mark: 1

   (b) the ratio of moles of reactant (gas): moles of product (gas) is 1:1 or the number of (gaseous) molecules is the same on both sides of the equation

   Maximum mark: 1

   (c) propene and cyclopropane curves both level off at the same concentrations as in graph on left hand page; ignore time axis

   Maximum mark: 1

12. (a)
$$CH_3-\underset{\underset{OH}{|}}{\overset{\overset{CH_3}{|}}{C}}-CH_2-\overset{\overset{CH_3}{|}}{C}=O$$

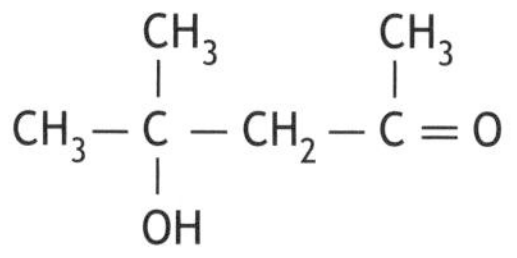

   Maximum mark: 1

   (b) methanal or formaldehyde

   Maximum mark: 1

   (c) water is not a product of the reaction or no small molecule produced or it is an addition reaction

   Maximum mark: 1

13. (a) $MnO_4^-(aq) + 8H^+(aq) + 5e^- \rightarrow Mn^{2+}(aq) + 4H_2O(\ell)$
   (state symbols not required)

   Maximum mark: 1

   (b) (i) first titre is a rough (or approximate) result or not accurate or an estimate or too far away from the others

   Maximum mark: 1

   (ii) Pipette

   Maximum mark: 1

   (iii) 5.81 (g) – 4 marks

   Correct use of 6:2 ratio to calculate moles of oxalic acid (1 mark)

   Correct scaling from 25 ➡ 500 (1 mark)

   Gfm of oxalic acid = 90 (1 mark)

   Calculation of a mass of oxalic acid (1 mark)

   Moles permanganate = $0.04 \times \dfrac{26.9}{1000} = 0.0010$

   Moles oxalic = $\dfrac{0.0010}{2} \times 6 = 0.0032$ (for 25cm³)

   Moles for 500cm³ = 0.065

   Mass oxalic = 0.065 × 90 = 5.81g

   Mass = 0.0538 × 90 = 4.84g

   Maximum mark: 4

14. (a) Octadec-9,12,15-trienoic acid

   (allow the interchange of hyphens and commas)

   Maximum mark: 1

   (b) (i) neutralisation

   Maximum mark: 1

   (ii) any mention that soaps have both hydrophobic/oil-soluble and hydrophilic/water-soluble parts (or alternative wording showing knowledge of these parts of the soap)

   Correct identification of the parts of this soap which dissolve in water and oil, COO^-/COONa/ONa^+ and the hydrophobic part of the molecule, the hydrocarbon chain

   Describe how this results in a 'ball-like' structure/globule (with the oil/grease held inside the ball) or micelle or mention of an emulsion.

   Maximum mark: 3

15. (a) x is O–H, y is C–H

   Maximum mark: 1

   (b) 2 peaks only: at 1705–1800 and 2800–3000

   Maximum mark: 1

HIGHER FOR CfE CHEMISTRY MODEL PAPER 2

Section 1

Question	Response
1.	C
2.	A
3.	C
4.	D
5.	C
6.	C
7.	A
8.	B
9.	B
10.	C
11.	D
12.	A
13.	A
14.	A
15.	C
16.	A
17.	A
18.	A
19.	B
20.	A

Section 2

1. (a) (i) Boron or Carbon or B or C or graphite or diamond

    Maximum mark: 1

    (ii) Monatomic

    Maximum mark: 1

    (iii) Lithium

    Maximum mark: 1

    (b) 32g

    1 mark for stoichiometric

    Understanding: 2 mol KNO_3 ➡ 1 mol O_2

    1 mark for showing conversion to grams and litres.
    e.g. 202.2 g →24 l

    1 mark is awarded for correct arithmetic calculation of mass need to produce 3.8 litres

    $\frac{3.8}{24} = 0.158$

    ½ mol O_2 ➡ 1 mol KNO_3
    0.158 ➡ 0.317 mol KNO_3
    Mass = 0.317 × 101.1 = 32g

    Maximum mark: 3

2. This is an open ended question.

    The whole candidate response should first be read to establish its overall quality in terms of accuracy and relevance to the problem/situation presented. There may be strengths and weaknesses in the candidate response: assessors should focus as far as possible on the strengths, taking account of weaknesses (errors or omissions) only where they detract from the overall answer in a significant way, which should then be taken into account when determining whether the response demonstrates reasonable, limited or no understanding. Assessors should use their professional judgement to apply the guidance below to the wide range of possible candidate responses.

    3 marks: The candidate has demonstrated a good conceptual understanding of the chemistry involved, providing a logically correct response to the problem/situation presented.

    This type of response might include a statement of principle(s) involved, a relationship or equation, and the application of these to respond to the problem/situation. This does not mean the answer has to be what might be termed an "excellent" answer or a "complete" one.

    2 marks: The candidate has demonstrated a reasonable understanding of the chemistry involved, showing that the problem/situation is understood.

    This type of response might make some statement(s) that is/are relevant to the problem/situation, for example, a statement of relevant principle(s) or identification of a relevant relationship or equation.

    1 mark: The candidate has demonstrated a limited understanding of the chemistry involved, showing that a little of the chemistry that is relevant to the problem/situation is understood.

    The candidate has made some statement(s) that is/are relevant to the problem/situation.

    0 marks: The candidate has demonstrated no understanding of the chemistry that is relevant to the problem/situation.

    The candidate has made no statement(s) that is/are relevant to the problem/situation.

    Points you could discuss:

    The definition of each of the terms.

    The basic structure of the atom and how the nuclear charge affects these properties.

    How electron shells cause screening.

    Illustrate your answers with illustrations and trends from the periodic table. Use the data booklet to quote values.

    Maximum mark: 3

3. (a) Experiment 2 curve initial gradient steeper than Experiment 1 and curve levels off at approximately same volume as Experiment 1 (1 mark)

    Experiment 3 curve initial gradient less steep than Experiment 1 and levels off at approximately half final volume of Experiment 1 (1 mark)

    Maximum mark: 2

ANSWERS TO HIGHER FOR CfE CHEMISTRY

(b) More molecules (particles) have enough energy to react

OR

more molecules have sufficient energy

OR

more molecules with (kinetic) energy greater than the activation energy

Maximum mark: 1

4. (a) −545 (kJ mol^{-1})

For 1 mark candidate must have (436 + 159) or 595 or (2 × 570) or 1140 or 545

Maximum mark: 2

(b) For 1 mark:

H-bonds between H-F molecules are stronger than LDF between F-F molecules

OR

for 1 mark – More energy is required to break H bonds in H$_2$ than LDF in F$_2$

For 1 mark: H bonds caused by:

(large) difference in electronegativity

OR

indication of polar bonds

OR

indication of permanent dipole

For 1 mark: LDF caused by: temporary dipoles

OR

uneven distribution of electrons

OR

electron cloud wobble

Maximum mark: 3

5. (a) Improve reliability/allow an average value to be calculated.

Maximum mark: 1

(b) 0·197 g

Partial marks can be awarded using a scheme of two "concept" marks and one "arithmetic" mark.

1 mark for knowledge of the relationship between moles, concentration and volume. This could be shown by any one of the following steps:

calculation of moles of iodine/vit C in sample
0·005 × 0·0224
calculation of moles of vit C in 500cm^3

1 mark for knowledge of the relationship between moles, mass and GFM of vit C ie 0.00112 × 176

1 mark is awarded for correct arithmetic throughout the calculation. This mark can only be awarded if both concept marks have been awarded.

$\left(\dfrac{22.4}{1000}\right) \times 0.0050 = 0.000112$.

Moles Vit C = 0.000112 for 50cm^3
Moles Vit C = 0.00112 for 500cm^3
Mass Vit C = 0.00112 × 176 = 0.197g

Maximum mark: 3

(c) 50cm^3 samples were not measured accurately or a measuring cylinder was used or a pipette should have been used

Maximum mark: 1

(d) (i) Hydrolysis

Maximum mark: 1

(ii) More OH (hydroxyl) or can form more hydrogen bonds

Maximum mark: 1

(e) 411 cm^3

7.3 mg ➡ 50 cm^3

1.0 mg ➡ $\dfrac{50}{7.3}$ = 6.85 cm^3

60 mg ➡ 60 × 6.85 = 411 cm^3

Maximum mark: 2

6. (a) (i) Tollen's or acidified dichromate or Fehling's or Benedict's (please note — although Benedict's reagent would not work in practice, because it appears in Higher textbooks, revision guides and the PPA materials for the traditional Higher, it can be accepted)

(accept other spellings if phonetically correct)

Maximum mark: 1

(ii) Carboxylic acid

Maximum mark: 1

(b) (i) It keeps oil & water soluble materials mixed

OR

Allow immiscible substances to mix

OR

To allow fat and water to mix

OR

To form a suspension

Zero: to stop it separating (with no mention of water & oil soluble components) to stop layers forming

Maximum mark: 1

(ii)
```
     H   H   H
     |   |   |
 H — C — C — C — H
     |   |   |
     OH  OH  OH
```

Maximum mark: 1

(c) 6·7 (mg) – units not required

A single mark is available if either of the following manipulations is correctly executed.

Correct use of percentage
eg mass of chocolate = 0.28 × 17 g = 4.76 g

Correct use of proportion theobromine
eg mass of theobromine = 1·4 × a mass

28% of 17g = 4.76g
1g ➡ 1.4mg
4.76g ➡ 6.66mg

Maximum mark: 2

(d) Open-ended question

Points you could discuss:
Differences in properties of ionic and covalent compounds (solubility in polar (water) and non-polar solvents; conductivity; melting points; volatility; typical flavouring compounds and their likely structure i.e. are they ionic or covalent; typical molecules you would expect to find in a food)

Maximum mark: 3

7. (a) (i) Carboxyl (group) (1 mark)
OR
Carboxylic (acid) (1 mark)
Maximum mark: 1

(ii) Structure: 2,6-dimethylphenyl group bonded to N–H (NH attached to benzene ring with CH₃ groups at ortho positions)

$$\text{CH}_3\text{-C}_6\text{H}_3(\text{CH}_3)\text{-NH-H}$$

Maximum mark: 1

(iii)
$$\text{Na}^+\text{O}-\overset{\displaystyle O}{\underset{\|}{C}}-\text{CH}_2-\text{N}\begin{array}{c}\text{CH}_2\text{CH}_3\\\text{CH}_2\text{CH}_3\end{array}$$

or shortened formula.

Charges not required but if shown, both +ve and –ve charges must be correct

Worth zero: Covalent bond shown between Na–O
Maximum mark: 1

(b) 25 (minutes)
OR
8.0 to 8.4 (minutes)
(units not required. Ignore incorrect units)
Maximum mark: 1

(c) Volume = 31·5 cm³ or 31·5 ml or 0·0315 l or equivalent (3 marks)

One mark is allocated to the correct statement of units of volume. This is the mark in the paper earmarked to reward a candidate's knowledge of chemical units.

So volume = 31·5 or 0·0315 (2 marks)

One mark is available if either of the following steps is correct

Calculation of mass of lidocaine

eg 4·5 × 70 = 315 (mg)

Calculation of a volume of solution required

3

eg a mass × 0.1 = a volume

70 × 4.5 = 315mg

$\frac{315}{10} \times 1 = 31.5$

Maximum mark: 3

(d) (i) Benzocaine is a smaller/Tetracaine is bigger
OR (1 mark)
weaker London Dispersion Forces with Benzocaine (1 mark)
OR
weaker Van der Waal's forces for Benzocaine
OR (1 mark)
Benzocaine has lower b.pt (1 mark)
OR
Benzocaine more soluble/attracted to/in mobile phase (1 mark)
OR
Benzocaine less strongly attracted to stationary phase (1 mark)
OR
Benzocaine is more polar (1 mark)

Worth zero: Benzocaine takes less time to travel through the apparatus
Maximum mark: 1

(ii) The peaks for lidocaine and caffeine overlap
OR
Candidate wording for idea of masking

Worth zero: The retention times are similar
Maximum mark: 1

(iii) Peak for tetracaine at correct RT with approximately half original height
Maximum mark: 1

8. (a) Any correct structural formula for CCl_2F_2 where each atom forms the correct number of bonds
Maximum mark: 1

(b) High energy radiation (UV) breaks bonds in molecules/excites electrons
Maximum mark: 1

(c) (i) Propagation
Maximum mark: 1

(ii) Free radical scavenger will react with Cl· (1 mark)

This will terminate free radical reactions (1 mark)
OR
This will prevent Cl· reacting with ozone (1 mark).
Maximum mark: 2

9. (a) $2B_2O_3 + 7C \rightarrow B_4C + 6CO$
Maximum mark: 1

(b) Diagram showing any workable method of producing CO_2 with calcium carbonate and dilute hydrochloric acid labelled (1 mark)
AND
removing CO_2 with chemical labelled, eg sodium hydroxide solution, lime water, alkali
Maximum mark: 2

(c) 46% or 0.46

1 mark for correct gfms
1 mark for correct use of formula

$$\frac{2 \times 56}{160 + (3 \times 28)} \times 100 = 46\%$$

10. (a) Correct identification of amide link (HN–C=O)
Maximum mark: 1

(b) (i) 51%

1 mark is given for either calculating the theoretical yield, or for working out the numbers of moles of reactant and product formed.
eg 6.23(g)

1 mark is given for calculating the % yield; either using the actual and theoretical masses, or using the actual number of moles of products and actual number of moles of reactant.

109g ➡ 151g

1 g ➡ $\frac{151}{109} = 1.39$

4.5g ➡ 4.5 × 1.39 = 6.23g (theoretical)

$\frac{3.2}{6.23} \times 100 = 51\%$

Maximum mark: 2

ANSWERS TO HIGHER FOR CfE CHEMISTRY

(ii) £40.22
703g – 1 mark

3.2g para ➡ 4.5g of 4AP

1g para ➡ $\frac{4.5}{3.2}$ = 1.41g

500g para requires 500 × 1.41 = 703g
1000g costs £57.20

$\frac{703}{1000}$ × 57.20 = £40.22

Maximum mark: 2

(c) Amino acids

Maximum mark: 1

(d) 0·0225 or 0·022 or 0·023

Maximum mark: 1

(e) Dissolve 0.040g l paracetamol (in a beaker) and transfer to a 1 litre standard flask with rinsings. Make up to the mark.

Maximum mark: 2

11. (a) Glycerol or propane-1,2,3-triol or propan-1,2,3-triol or glycerin

Maximum mark: 1

(b) Fat molecules have fewer/no double bonds/more saturated
OR
oil molecules have more double bonds/unsaturated (or similar)

Maximum mark: 1

(c) Oxygen

Maximum mark: 1

12. (a) Denaturing / denature

Maximum mark: 1

(b) (i) To prevent the temperature rising too high or gentle method of heating or to prevent the protein structure changing or to prevent denaturing of protein or to prevent separation of protein and fat or mention of flammability

Maximum mark: 1

(ii) Correct drawing of any one of the three amino acids showing -NH$_2$ and -COOH groups complete

Maximum mark: 1

(iii) [Structural formulae of trisodium salt of citric acid shown in two forms with -O-Na and -O$^-$ Na$^+$ groups]

Maximum mark: 1

13. (a) –803, 726, 283

(any two values from this list) (1 mark)
+206 kJ mol^{-1}
(for value, no follow through, units not required)
(1 mark)

Maximum mark: 2

(b)

| temperature | (decrease) / keep the same / increase |
| pressure | decrease / keep the same / (increase) |

Maximum mark: 1

(c) –567 kJ mol^{-1}

Partial marks can be awarded using a scheme of two "concept" marks and one "arithmetic" mark.

1 mark for demonstration of use of the relationship Eh = cmΔT this mark is for the concept, do not penalise for incorrect units or incorrect arithmetic.

The value of 6.7 (kJ) would automatically gain this mark.

1 mark for demonstration of knowledge that the enthalpy value provided relates to 1 mole of methanol. This mark could be awarded if the candidate's working shows a proportion calculation involving use of the gfm for methanol (32).

1 mark is awarded for correct arithmetic throughout the calculation. This mark can only be awarded if both concept marks have been awarded.

E = 4.18 × 0.1 × 16.1 = 6.73 kJ
0.38g ➡ 6.73 kJ

$\frac{6.73}{0.38}$ = 17.71 kJ

32g ➡ 32 × 17.71 = 567 kJ

Maximum mark: 3

(d) complete combustion (or incomplete combustion in lab method) or richer supply of oxygen (or burns in air in lab method) or no evaporation of methanol

Maximum mark: 1

14. (i) left to right 3, 2, 1

Maximum mark: 1

(ii) but-2-ene

Maximum mark: 1

HIGHER FOR CfE CHEMISTRY MODEL PAPER 3

Section 1

Question	Response
1.	C
2.	D
3.	B
4.	A
5.	D
6.	B
7.	B
8.	C
9.	B
10.	A
11.	C
12.	D
13.	B
14.	C
15.	D
16.	D
17.	B
18.	D
19.	C
20.	B

Section 2

1. (a) Bonding – metallic and covalent

    Structure – network and molecular

    Maximum mark: 2 – all correct
    Maximum mark: 1 – two or more correct

    (b) Increasing nuclear charge
    OR
    Stronger nuclear pull

    Maximum mark: 1

    (c) (i) (Fractional) distillation

    Maximum mark: 1

    (ii) 5.8×10^7 tonnes
    OR
    57692307 tonnes

    1.3% of air = 750000 tonnes
    0.013 x = 750000 tonnes

    $x = \dfrac{750000}{0.013} = 57692307$

    (iii) Magnesium oxide would form
    OR
    magnesium would react with oxygen
    OR
    magnesium would react with nitrogen

    Maximum mark: 1

    (iv) This is an open ended question.

    The whole candidate response should first be read to establish its overall quality in terms of accuracy and relevance to the problem/situation presented. There may be strengths and weaknesses in the candidate response: assessors should focus as far as possible on the strengths, taking account of weaknesses (errors or omissions) only where they detract from the overall answer in a significant way, which should then be taken into account when determining whether the response demonstrates reasonable, limited or no understanding. Assessors should use their professional judgement to apply the guidance below to the wide range of possible candidate responses.

    3 marks: The candidate has demonstrated a good conceptual understanding of the chemistry involved, providing a logically correct response to the problem/situation presented

    This type of response might include a statement of principle(s) involved, a relationship or equation, and the application of these to respond to the problem/situation. This does not mean the answer has to be what might be termed an "excellent" answer or a "complete" one.

    2 marks: The candidate has demonstrated a reasonable understanding of the chemistry involved, showing that the problem/situation is understood.

    This type of response might make some statement(s) that is/are relevant to the problem/situation, for example, a statement of relevant principle(s) or identification of a relevant relationship or equation

    1 mark: The candidate has demonstrated a limited understanding of the chemistry involved, showing that a little of the chemistry that is relevant to the problem/situation is understood.

    The candidate has made some statement(s) that is/are relevant to the problem/situation.

    0 marks: The candidate has demonstrated no understanding of the chemistry that is relevant to the problem/situation.

    The candidate has made no statement(s) that is/are relevant to the problem/situation.

    Points you could discuss:

    The "nitrogen" prepared from method 1 would not be pure nitrogen-what else would it contain? Compare the volume produced in method 1 with method 2. If Mg was heated with the sample from method 1, how would this help?

    Maximum mark: 3

2. (a) (i) H — Cl + H•

   H$_2$

   1 mark for both species

   Maximum mark: 1

   (ii) to prevent light/UV shining on sample

   OR

   to prevent initiation

   OR

   to prevent radicals from forming

   OR

   to prevent shattering

   OR

   to prevent premature explosion

   Maximum mark: 1

   (b) For 1 mark candidate can have:
   "436+243" or "679" or "2×432" or "864" or "185"

   For 2 marks must have: −185

   Maximum mark: 2

3. (a) Palm oil has lower degree of unsaturation/palm oil less unsaturated/palm oil more saturated/palm oil contains more saturates/fewer double bounds

   OR

   Molecules in palm oil can pack more closely together

   Maximum mark: 1

   (b) Polyunsaturated

   Maximum mark: 1

   (c) (i) Soap/emulsifying agent/detergent/washing/cleaning

   Maximum mark: 1

   (ii) Glycerol or propane-1,2,3-triol

   Maximum mark: 1

   (d) Oxygen

   Maximum mark: 1

4 (a) Gas syringe

   Maximum mark: 1

   (b) (i) Reacting particles have more energy

   Maximum mark: 1

   (ii) Hydrogen gas is flammable/explosive

   Maximum mark: 1

   (c) 455g

   Partial marks

   15mg (1 mark)

   22% ➡ 3.3mg

   100% ➡ 15mg

   3.3mg ➡ 100g peanuts

   $\frac{100}{3.3}$ = 30.3g

   15mg ➡ 455g

5. (a) Ratio of oxygen:hydrogen atoms increased (or ratio of hydrogen:oxygen atoms decreased) or removal of hydrogen

   loss of electrons–zero

   Maximum mark: 1

(b) (i) Distinctive (fruity) smell

   OR

   insoluble/oily layer

   Maximum mark: 1

   (ii)

   $$CH_3-CH_2-\overset{\overset{O}{\|}}{C}-O-CH_2-CH_2-CH_3$$

   Maximum mark: 1

6. (a) One we need to get in the food we eat (from our diet) or one that the body cannot manufacture (make)

   Maximum mark: 1

   (b) Peptide link correctly identified

   $$-\overset{\overset{O}{\|}}{C}-\overset{\overset{H}{|}}{N}-$$

   Maximum mark: 1

   (c)

   $$H_2N-\overset{\overset{CH_3}{|}}{\underset{\underset{H}{|}}{C}}-\overset{\overset{O}{\|}}{C}-\overset{\overset{H}{|}}{N}-\overset{\overset{CH_2-\text{Ph}}{|}}{\underset{\underset{H}{|}}{C}}-\overset{\overset{O}{\|}}{C}-OH$$

   Maximum mark: 1

7. (a) (i) Compare this chromatogram with the chromatogram from a genuine sample

   Maximum mark: 1

   (ii) Methanol is more polar or ethanal is less polar

   Methanol is smaller-zero

   Maximum mark: 1

   (iii) Any time between 5.5 and 9.5mins

   Maximum mark: 1

   (iv) The peaks overlap OR the retention times are too close

   Maximum mark: 1

   (b) (i) C=O circled

   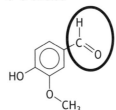

   Maximum mark: 1

   (ii) Tollen's or Fehling's or acidified dicromate

   Maximum mark: 1

   (iii) Line drawn from hydroxyl (−OH) H of furaneol to O of water OR from any O of furaneol to H of water

   Maximum mark: 1

   (c) Open-ended question

   Points you could discuss:

   How might the ethanol content change through evaporation or oxidation? What could it change to? Could the ethanol react with other compounds? What about the other compounds-could they oxidise? Evaporate? React with something else?

   Maximum mark: 3

8. (a) 0.51 litres or 510 cm³

   1 mark for stoichiometric

   Understanding: 1 mol Cu(NO₃)₂ → 2 mol NO₂

   1 mark for showing conversion of grams to litres.
   187.5 g → 48 l

   1 mark is awarded for correct arithmetic calculation of volume produced by 2g.

   2·0 g → 2·0/187.5 × 48 = 0.51 litres

   1 mol Cu(NO₃)₂ → 2 mol NO₂

   187.5 g → 2×24= 48 litres

   $\frac{48}{187.5}$ = 0.256 litres

   2g → 2× 0.256 = 0.51 litres

   Maximum mark: 1

   (b) Diagram showing any method of condensing the nitrogen dioxide and workable way of collection eg U tube in ice

   Diagram showing gas bubbling through water-zero

   Maximum mark: 1

9. (a) 2NaOH + Cl₂ → NaClO + NaCl + H₂O – 2 marks

   Correct formulae for all but one substance, whether balanced or not – 1 mark

   Maximum mark: 2

   (b) NaOH results in decrease in H⁺ ion concentration – 1 mark

   Position of equilibrium moves to the right and ClO increases — 1 mark

   hypochlorite bleach is made by reacting sodium hydroxide with chlorine – 0 marks

   (c) The electrons can be shown as 2e without penalty
   ClO⁻ + 2H⁺ + 2e → Cl⁻ + H₂O

   (d) 0·67 mol l⁻¹ – 4 marks

   Partial marks can be awarded using a scheme of two "concept" marks, one "arithmetic" mark and one "unit" mark.

   1 mark – correct strategy to work out number of moles of oxygen (a volume of oxygen in whatever unit is shown being divided by a molar volume in whatever unit)

   1 mark is awarded for correct strategy to work out the concentration of bleach (eg a number of moles of ClO is divided by the volume of the sample in any units)

   1 mark is awarded for correct arithmetic in both of these steps- units correct within both calculations (ie. Gas volume and molar volume both in cm³ or both in l, volume in concentration calculation in l) This arithmetic mark can only be awarded if both of the concept marks are awarded.

   1 mark is allocated to the correct statement of units of concentration.

   This is the mark in the paper earmarked to reward a candidate's knowledge of chemical units.

   Moles Oxygen = $\frac{80}{24000}$
   = 0.0033 mol

   Conc = $\frac{mol}{vol}$ = $\frac{0.0033}{0.005}$ = 0.67 moll⁻¹

   Maximum mark: 1

10. (a)

    CH₃ — CH₂ — CH — CH₃
                  |
                  OH

    Maximum mark: 1

    (b) For 1 mark

    H-bonds between triethanol amine molecules are stronger than LDF between triisopropyl amine molecules

    OR

    for 1 mark – More energy is required to break H bonds in triethanol amine than LDF in triisopropyl amine

    For 1 mark

    H bonds caused by:

    (large) difference in electronegativity

    OR

    indication of polar bonds

    OR

    indication of permanent dipole

    For 1 mark

    LDF caused by:
    temporary dipoles

    OR

    uneven distribution of electrons

    OR

    electron cloud wobble

    Maximum mark: 3

11. (a) 75 % (2)

    Total mass of reactants/products = 240 g (1)

    Atom Economy $\frac{180}{240}$ × 100% = 75% (1)

    (Accept 0·75 also, this would be atom economy as a fraction

    Maximum mark: 2

    (b) 40% (2)

    1 mark is given for either calculating the theoretical yield, or for working out the numbers of moles of reactant and product formed.

    eg 6.55(g) or both 0.0364 and 0.0146

    1 mark is given for calculating the % yield; either using the actual and theoretical masses, or using the actual number of moles of products and actual number of moles of reactant.

    138g → 180g

    1g → $\frac{180}{138}$ = 1.304

    5.02 → 5.02 × 1.304 = 6.55g

    % yield = $\frac{2.62}{6.55}$ × 100 = 40%

    Maximum mark: 2

12. (a) (i) 5.75 or 5.77 (g – units not required, ignore incorrect units)

    Partial marks can be awarded using a scheme of two "concept" marks, and one "arithmetic" mark

    1 mark – for demonstration of use of the relationship between specific heat capacity, mass, temperature and heat energy/enthalpyeg
    Eh = cmΔT

    This mark is for the concept, do not penalise for incorrect units or incorrect arithmetic.

    The value of 171 (kJ) would automatically gain this mark.

    1 mark – for demonstration of knowledge that the enthalpy of combustion of ethanol relates to the combustion of the gfm of ethanol. This mark could be awarded if the candidate is seen to be working out the number of moles of ethanol required (0.125 or 0.13) or if the candidates working shows a proportion calculation involving use of the gfm for ethanol (46). This mark is for demonstration of knowledge of this concept, do not penalise for incorrect units of incorrect arithmetic.

    1 mark – the final mark is awarded for correct arithmetic throughout the calculation but cannot be awarded unless the two concept marks have both been awarded

    E = 4.18 × 0.5 × 82 = 171kJ

    1367 ⟶ 46g

    1 ⟶ $\frac{1376}{46}$ = 0.034

    171 ⟶ 171 × 0.034 = 5.75g

    Maximum mark: 3

    (ii) Heat lost to surroundings (1 mark)

    Incomplete combustion (of alcohol) (1 mark)

    Ethanol impure (1 mark)

    Loss (of ethanol) through evaporation (1 mark)

    Maximum mark: 1

    (b) 1 660 000 (kJ – units not required, ignore incorrect units)

    Partial marks 1 mark for ratio or 34500 or 34.5 appearing in working

    $1.45cm^3$ ⟶ 1g

    $1cm^3$ ⟶ 0.69g

    $50000cm^3$ ⟶ 34482g

    E = 34482 × 48 = 1655172kJ

    Maximum mark: 2

13. (a) (i) 3-methyl-butan-2-ol
    (with or without the hyphens)

    Maximum mark: 1

    (ii) $CH_3—CH_2—CH_2—\underset{\underset{H}{|}}{\overset{\overset{CH_3}{|}}{C}}—\underset{\underset{OH}{|}}{\overset{\overset{H}{|}}{C}}—H$

    Any correct structural formula for 2-methylpentan-1-ol

    Maximum mark: 1

    (b) –2168 (kJ mol^{-1})

    1 mark for two from the three correct enthalpy change values: –36 kJ   –1274 kJ   3 × –286 (= –858) kJ

    +2168 without any working is worth 0

    Maximum mark: 2

14. (i) First titre is a rough (or approximate) result/ practice
    OR
    first titre is not accurate/not reliable/rogue
    OR
    first titre is too far away from the others
    OR
    you take average of concordant/close results

    Maximum mark: 1

    (ii) 0·045 (mol l^{-1} – units not required) 3 marks

    Partial marks can be awarded using a scheme of two "concept" marks, and one "arithmetic" mark.

    1 mark for knowledge of the relationship between moles, concentration and volume. This could be shown by any one of the following steps:

    0·1 × 0·01815 = 0·001815

    (0·000908) ÷ 0·02 = 0·045

    Insertion of correct pairings of values for concentration and volume in a valid titration formula

    1 mark for knowledge of relationship between moles of thiosulfate and iodine. This could be shown by any one of the following steps:

    0·001815/2 = 0·000908

    Insertion of correct stoichiometric values in a valid titration formula

    1 mark is awarded for correct arithmetic throughout the calculation. This mark can only be awarded if both concept marks have been awarded.

    Moles = 0.01815 × 0.1 = 0.001815

    0.001815/2 = 0.0009075

    0.0009075/0.02 = 0.045 $moll^{-1}$

    Maximum mark: 3

    (iii) Mass of sodium thiosulfate = 3·96 g (1 mark)

    Mention of rinsings (1 mark)

    Mention of make up to the mark (1 mark)

    Maximum mark: 3

15. (a) 3

    Maximum mark: 1

    (b) 0.204 °C Also accept 0.02 °C

    Maximum mark: 1

HIGHER FOR CfE CHEMISTRY 2015

Section 1

Question	Response
1.	D
2.	C
3.	D
4.	A
5.	C
6.	B
7.	A
8.	B
9.	B
10.	A
11.	D
12.	D
13.	B
14.	C
15.	B
16.	A
17.	D
18.	A
19.	A
20.	C

Section 2

1. (a) Sulfur — London dispersion forces/van der Waals/ intermolecular forces (1 mark)

    Silicon dioxide — covalent **or** polar covalent **or** covalent network bonds

    (1 mark)

    Maximum mark: 2

   (b) (i) Any structure for P_4S_3 that obeys valency rules

   Maximum mark: 1

   (ii) Sulfur has more protons in nucleus **or** sulfur has increased nuclear pull for electrons **or** increased nuclear charge

   Maximum mark: 1

   (iii) Correctly identify that the forces are stronger between sulfur (molecules) than between the phosphorus molecules (1 mark)

   Correctly identifying that there are London dispersion forces between the molecules of both these elements (1 mark)

   These forces are stronger due to sulfur structure being S8 whereas phosphorus is P4 (1 mark)

   Maximum mark: 3

2. (a) From graph, rate = 0·022

   t = 1/rate = 45s

   Maximum mark: 1

   (b) (i) Second line displaced to left of original. Peak of curve should be to the left of the original peak

   (ii) A vertical line drawn at a lower kinetic energy that the original Ea shown on graph

   Maximum mark: 1

3. (a) (i) Workable apparatus for passing steam through strawberry gum leaves. Steam must pass **through** the strawberry gum leaves (1 mark)

   Workable apparatus for condensing the steam and essential oil (1 mark)

   Maximum mark: 2

   (ii) (Fractional) distillation or chromatography

   Maximum mark: 1

   (b) (i) **1 mark** awarded for correct arithmetical calculation of moles of acid = 0·044 and moles alcohol = 0·063

   OR

   working out that 9·25g cinnamic acid would be needed to react with 2g methanol or 6·5g cinnamic acid would react with 1·41g methanol

   1 mark awarded for statement demonstrating understanding of limiting reactant.

   e.g. there are less moles of cinnamic acid therefore it is the limiting reactant

   OR

   0·0625 moles methanol would require 0·0625 moles cinnamic acid

   OR

   methanol is in excess therefore cinnamic acid is the limiting reactant.

   Maximum mark: 2

   (ii) (A) 52%

   1 mark is given for working out the theoretical yield ie 7·1g

   OR

   for working out both the moles of reactant used **AND** product formed ie both 0·044 moles and 0·023 moles

   1 mark is given for calculating the % yield, either using the actual and theoretical masses

   OR

   using the actual number of moles of products and actual number of moles of reactant

   Maximum mark: 2

   (B) £24·59

   1 mark for

   Evidence for costing to produce of 3·7g (£0·91)

   OR

   evidence of a calculated mass of cinnamic acid × 14p

   OR

   evidence that 176g of cinnamic acid required £12·80 would be using 100% yield

   Maximum mark: 2

4. (a) Any one of the common compounds correctly identified i.e.

   citronellol/geraniol/anisyl alcohol
   Maximum mark: 1

   (b) Compounds that are common to the brand name and counterfeit perfumes are present in lower concentration in the counterfeit perfume

   OR

   Smaller volumes of compounds that are common to the brand name and counterfeit perfumes are used in the counterfeit perfume
   Maximum mark: 1

   (c) (i) Inert/will not react with the molecules (being carried through the column)
   Maximum mark: 1

   (ii) Size (mass) of molecules / temperature of column.
   Maximum mark: 1

   (d) (i) Terpenes
   Maximum mark: 1

   (ii) (A) 3,7-dimethylocta-1,6-dien-3-ol
   Maximum mark: 1

   (B) Hydroxyl attached to C which is attached to 3 other C atoms

   OR

   hydroxyl attached to a C that has no H atoms attached
   Maximum mark: 1

   (e) 1·7 g (units not required)

5. This is an open ended question

   1 mark: The student has demonstrated, at an appropriate level, a limited understanding of the chemistry involved. The student has made some statement(s) which is/are relevant to the situation, showing that at least a little of the chemistry within the problem is understood.

   2 marks: The student has demonstrated a reasonable understanding, at an appropriate level, of the chemistry involved. The student makes some statement(s) which is/are relevant to the situation, showing that the problem is understood.

   3 marks: The maximum available mark would be awarded to a student who has demonstrated a good understanding, at an appropriate level, of the chemistry involved. The student shows a good comprehension of the chemistry of the situation and has provided a logically correct answer to the question posed. This type of response might include a statement of the principles involved, a relationship or an equation, and the application of these to respond to the problem. This does not mean the answer has to be what might be termed an "excellent" answer or a "complete" one.

   Points that could have been discussed include:
   trends in Periodic properties such as covalent radius, electronegativity and ionisation energies of the elements in the groups; bonding and structures of the elements in the groups.
   Maximum mark: 3

6. (a) Heat breaks hydrogen bonds
   Maximum mark: 1

   (b) (i) Either of structures shown circled

   Maximum mark: 1

   (b) (ii) 50·5 ± 1 °C
   Maximum mark: 1

   (c) (i) Hydrolysis
   Maximum mark: 1

   (ii) (A) 5
   Maximum mark: 1

   (B) Correctly drawn amino acid structure

```
    H   H   O
    |   |   ||
H — N — C — C — OH
        |
        CH₂
        |
        CH₂
        |
        C = O
        |
        OH
```
$$\text{Maximum mark: 1}$$

7. (a) 118/32

    OR

    3·69 mol CH3OH (1 mark)

    3·69 × 24 = 88·5 litres (1 mark)

    Maximum mark: 2

   (b) (i) (A) Thermometer touching bottom or directly above flame

       OR

       temperature rise recorded would be greater than expected

       Maximum mark: 1

       (B) Distance between flame and beaker

       OR

       Height of wick in burner

       Same type of beaker

       Same draught proofing

       Maximum mark: 1

       (C) 2 concept marks and 1 arithmetic mark

       Concept marks

       Demonstration of the correct use of the relationship $E_h = cm\Delta T$ (1 mark) e.g. 4·18 × 0·1 × 23 OR 9·61

       AND

       Knowledge that enthalpy of combustion relates to 1 mol (1 mark) evidenced by scaling up of energy released

       Correct arithmetic = −288 kJ mol⁻¹ (1 mark)

       Maximum mark: 3

       (ii) 0·799 (0·8)

       Maximum mark: 1

   (c) (i) If reactions are exothermic heat will need to be removed/If reactions are endothermic heat will need to be supplied

       OR

       Chemists can create conditions to maximise yield

       Maximum mark: 1

       (ii) Answer = +191 kJ mol⁻¹ (2)

       Evidence of the use of all the correct bond enthalpies (1 mark) (412, 360, 463, 436, 743)

       OR

       Correct use of incorrect bond enthalpy values

       Maximum mark: 2

8. (a) Calcium carbonate/carbon dioxide/ammonia/calcium oxide all correctly identified in flow diagram (1 mark)

    Ammonium chloride/sodium hydrogen carbonate/sodium carbonate/water — all correctly identified in flow diagram (1 mark)

    Maximum mark: 2

   (b) (Adding brine) increases sodium ion concentration hence equilibrium shifts to right (1 mark)

    Rate of forward reaction is increased (by addition of brine) (1 mark)

    Maximum mark: 2

9. This is an open ended question

    1 mark: The student has demonstrated, at an appropriate level, a limited understanding of the chemistry involved. The candidate has made some statement(s) at which is/are relevant to the situation, showing that at least a little of the chemistry within the problem is understood.

    2 marks: The student has demonstrated, at an appropriate level, a reasonable understanding of the chemistry involved. The student makes some statement(s) which is/are relevant to the situation, showing that the problem is understood.

    3 marks: The maximum available mark would be awarded to a student who has demonstrated, at an appropriate level, a good understanding of the chemistry involved. The student shows a good comprehension of the chemistry of the situation and has provided a logically correct answer to the question posed. This type of response might include a statement of the principles involved, a relationship or an equation, and the application of these to respond to the problem. This does not mean the answer has to be what might be termed an "excellent" answer or a "complete" one.

    Points that could have been discussed include: nature of crude oil and vegetable oil; cleaning action of detergents; emulsions.

    Maximum mark: 3

10. (a) (i) 24 hours allows time for all of the zinc to react (1 mark)

    No stopper allows hydrogen gas to escape from the flask (1 mark)

    Maximum mark: 2

    (ii) Zinc ions/impurities/metal ions/salts may be present in tap water

    (b) (i) Pipette

    Maximum mark: 1

    (ii) 10 (Units not required, if given mg per litre, mg l⁻¹)

    Maximum mark: 1

    (c) Answer in range 4·6—4·8

    (mg per litre, mg l⁻¹)

    Maximum mark: 1

11. (a) Carboxyl/carboxylic (acid) group

    Maximum mark: 1

    (b) Esterification/condensation

    Maximum mark: 1

    (c)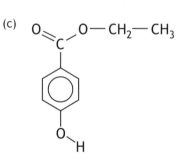

    Maximum mark: 1

(d) As molecular size (no. of carbon atoms) increases, the absorption decreases

Maximum mark: 1

12. (a) (i) 3 points

    1 mark for rinsing the burette — rinse the burette with the thiosulfate/required solution/ with the solution to be put in it.

    2 marks (1 mark each) for any 2 of the following points

    - fill burette above the scale with thiosulfate solution
    - filter funnel used should be removed
    - tap opened/some solution drained to ensure no air bubbles
    - (thiosulfate) solution run into scale
    - reading should be made from bottom of meniscus

    Maximum mark: 3

    (ii) $2I^-(aq) \rightarrow I_2(aq) + 2e^-$

    Maximum mark: 1

    (iii) 0·000062 (mol l^{-1})

    Scheme of two "concept" marks, and one "arithmetic" mark

    1 mark for knowledge of the relationship between moles, concentration and volume. This could be shown by one of the following steps:

    Calculation of moles thiosulfate solution e.g. 0·001 × 0·0124 = 0·0000124

    OR

    calculation of concentration of iodine solution e.g. 0·0000062/0·1

    OR

    Insertion of correct pairings of values for concentration and volume in a valid titration formula

    1 mark for knowledge of relationship between moles of thiosulfate and hypochlorite. This could be shown by one of the following steps:

    Calculation of moles hypochlorite from moles thiosulfate — e.g. 0·0000124/2 = 0·0000062

    OR

    Insertion of correct stoichiometric values in a valid titration formula

    1 mark is awarded for correct arithmetic through the calculation. This mark can only be awarded if both concept marks have been awarded.

    Maximum mark: 3

(b) **1 mark** correct arithmetic, either 44·4 (litres) or 44400 (cm^3)

   1 mark correct units

   Maximum mark: 2

(c) (i) **1 mark** Ammonia is polar and trichloramine is non-polar

   1 mark Explanation of this in terms of polarities of bonds **OR** electronegativity differences of atoms in bonds

   Maximum mark: 2

   (ii) Substances that have unpaired electrons

   Maximum mark: 1

   (iii) Propagation

   Maximum mark: 1

13. (a) Aldehyde group correctly identified

   Maximum mark: 1

(b) Ring form correctly drawn

   Maximum mark: 1

Acknowledgements

Permission has been sought from all relevant copyright holders and Hodder Gibson is grateful for the use of the following:

An extract from 'Patterns in the Periodic Table' © Royal Society of Chemistry (2015 Section 2 page 14).

Hodder Gibson would like to thank SQA for use of any past exam questions that may have been used in model papers, whether amended or in original form.